Enterprise DevOps for Architects

Leverage AIOps and DevSecOps for secure
digital transformation

Jeroen Mulder

BIRMINGHAM—MUMBAI

Enterprise DevOps for Architects

Group Product Manager: Rahul Nair

Publishing Product Manager: Niranjan Naikwadi

Senior Editor: Shazeen Iqbal

Content Development Editor: Romy Dias

Technical Editor: Arjun Varma

Copy Editor: Safis Editing

Project Coordinator: Shagun Saini

Proofreader: Safis Editing

Indexer: Manju Arasan

Production Designer: Roshan Kawale

First published: September 2021

Production reference: 1010921

Published by Packt Publishing Ltd.
Livery Place
35 Livery Street
Birmingham
B3 2PB, UK.

ISBN 978-1-80181-215-3

www.packt.com

To everyone in this world who fights for inclusion and diversity, making this world a place where every person may be whoever they choose to be.

Contributors

About the author

Jeroen Mulder (born 1970) started his career as an editor for Dutch newspapers. His IT career spans over 23 years, starting at Origin. Origin transformed into Atos Origin and eventually Atos, where Jeroen fulfilled many roles, lastly as a principal architect. In 2017, he joined Fujitsu as the lead architect, specializing in cloud technology. From June 2021, he has been working as the principal cloud solution architect at Philips, Precision Diagnosis.

Jeroen is a certified enterprise and security architect, concentrating on cloud technology. This includes architecture for cloud infrastructure, serverless and container technology, application development, and digital transformation using various DevOps methodologies and tools.

I want to thank my wonderful wife, Judith, and daughters, Rosalie and Noa, for all their support. I'd also like to thank my respective employers, Fujitsu and Philips, for granting me the opportunity and time to write this second book. And again, a big thank you goes to my editors at Packt Publishing, especially to Romy Dias, who edited most of my work.

About the reviewer

Werner Dijkerman is a DevOps engineer currently working for Fullstaq, specialists in Linux, DevOps, DataOps, Kubernetes, observability, the cloud, and cloud-native. He mostly does interim engineering for enterprise customers. He is currently focused and working on cloud-native solutions and tools such as Azure, AWS, Kubernetes, and Terraform. His focus is on Infrastructure as Code, monitoring to correct *things*, automating everything, and preventing from having to do anything that resembles manual work.

He was also a technical reviewer for the second and third editions of the *Zabbix Network Monitoring* book, *Learning Docker, Second Edition*, and various Packt video courses such as *Docker: Tips, Tricks, and Techniques* and *Kubernetes in 7 Days*.

Big thanks, hugs, and a shout out to Judith Borgers, Marije Titulaer, Fabian Met, Gerrit Tamboer, Guston Remie, and the rest of the team at Fullstaq!

Table of Contents

3
Architecting for DevOps Quality

4
Scaling DevOps

5
Architecting Next-Level DevOps with SRE

Section 2: Creating the Shift Left with AIOps

6

Defining Operations in Architecture

7

Understanding the Impact of AI on DevOps

8

Architecting AIOps

9

Integrating AIOps in DevOps

10

Making the Final Step to NoOps

Section 3: Bridging Security with DevSecOps

11

Understanding Security in DevOps

12

Architecting for DevSecOps

13

Working with DevSecOps Using Industry Security Frameworks

14

Integrating DevSecOps with DevOps

15

Implementing Zero Trust Architecture

Assessments

Other Books You May Enjoy

Index

Preface

This book provides an architectural overview of DevOps, AIOps, and DevSecOps – three domains that drive and accelerate digital transformation. Complete with step-by-step explanations of essential concepts, practical examples, and self-assessment questions, you will understand why and how DevOps should be integrated into the enterprise architecture, securing it and resulting in leaner operations and accelerating digital transformation.

By the end of the book, you will be able to develop robust DevOps architectures and understand which toolsets you can use. You will also have a deeper understanding of next-level DevOps implementing Site Reliability Engineering. You will have learned what AIOps is and what value it can bring to the enterprise. Lastly, you will have learned how to integrate security principles such as Zero Trust and industry security frameworks to DevOps with DevSecOps.

Who this book is for

This book teaches enterprise architects, solutions architects, and consultants how to design, implement, and integrate DevOps in the enterprise architecture while keeping operations on board with agile development and with the enterprise remaining secure and resilient. The book is not about hundreds of different tools but will teach you how to create a successful DevOps architecture.

What this book covers

Chapter 1, *Defining the Reference Architecture for Enterprise DevOps*, introduces DevOps as part of the enterprise architecture.

Chapter 2, *Managing DevOps from Architecture*, shows how to manage DevOps artifacts such as Continuous Integration and Continuous Development from architecture.

Chapter 3, *Architecting for DevOps Quality*, teaches you how to ensure and manage quality attributes in DevOps, focusing on testing methodologies.

Chapter 4, Scaling DevOps, demonstrates that DevOps is not a one-time exercise but needs to scale with the enterprise's needs.

Chapter 5, Architecting Next-Level DevOps with SRE, introduces the concept of **Site Reliability Engineering** (**SRE**) and how an architect can prepare for SRE.

Chapter 6, Defining Operations in Architecture, focuses on the role of operations in enterprises and how this role is affected by DevOps.

Chapter 7, Understanding the Impact of AI on DevOps, elaborates on the impact of implementing **artificial intelligence** (**AI**) and **machine learning** (**ML**) on operations.

Chapter 8, Architecting AIOps, runs through the architectural steps to implement AI and ML in operations.

Chapter 9, Integrating AIOps in DevOps, discusses integrating AIOps into **Continuous Integration and Continuous Deployment** (**CI/CD**) and automating the CI/CD pipeline using AIOps.

Chapter 10, Making the Final Step to NoOps, is a bit of a philosophical chapter on the evolution of operations in an enterprise, starting from operations that are enhanced with AI and ML to NoOps, fully automated operations that don't require any manual intervention.

Chapter 11, Understanding Security in DevOps, introduces the security concepts that apply to DevOps, such as scanning code for vulnerabilities.

Chapter 12, Architecting for DevSecOps, discusses creating a security architecture that can be integrated into development and deployment pipelines.

Chapter 13, Working with DevSecOps Using Industry Security Frameworks, elaborates on various security frameworks, such as ISO and CSA, but also HIPAA for healthcare and PCI for financial institutions, which enterprises need to be compliant with in their DevOps practice.

Chapter 14, Integrating DevSecOps with DevOps, teaches you how to integrate security policies, standards, and guardrails in DevOps practices and how to govern DevSecOps.

Chapter 15, Implementing Zero Trust Architecture, is the final chapter on applying and managing Zero Trust in DevOps.

To get the most out of this book

It's highly recommended that you have a basic understanding of enterprise and cloud architecture. For enterprise architecture, an introduction to **The Open Group Architecture Framework (TOGAF)** is advised. For cloud architecture, a basic understanding of major public clouds such as AWS and Azure will certainly help you to get the most out of this book.

Download the color images

We also provide a PDF file that has color images of the screenshots and diagrams used in this book. You can download it here: `https://static.packt-cdn.com/downloads/9781801812153_ColorImages.pdf`.

Conventions used

Tip or important notes throughout the book will appear as follows:

> **Tips or important notes**
> Appear like this.

Get in touch

Feedback from our readers is always welcome.

General feedback: If you have questions about any aspect of this book, email us at `customercare@packtpub.com` and mention the book title in the subject of your message.

Errata: Although we have taken every care to ensure the accuracy of our content, mistakes do happen. If you have found a mistake in this book, we would be grateful if you would report this to us. Please visit `www.packtpub.com/support/errata` and fill in the form.

Piracy: If you come across any illegal copies of our works in any form on the internet, we would be grateful if you would provide us with the location address or website name. Please contact us at `copyright@packt.com` with a link to the material.

If you are interested in becoming an author: If there is a topic that you have expertise in and you are interested in either writing or contributing to a book, please visit `authors.packtpub.com`.

Share Your Thoughts

Once you've read *Enterprise DevOps for Architects*, we'd love to hear your thoughts!
Scan the QR code below to go straight to the Amazon review page for this book and
share your feedback.

https://packt.link/r/1801812152

Your review is important to us and the tech community and will help us make sure we're
delivering excellent quality content.

Section 1: Architecting DevOps for Enterprises

The objective of this first part is to give guidelines and guardrails to help you develop an architecture for DevOps within enterprises. After completion, you will be able to define a DevOps architecture that is aligned with enterprise architecture. You will be able to work with (business) **service-level agreements (SLAs)** and **key performance indicators (KPIs)** in DevOps components and work according to the **VOICE (Value, Objectives, Indicators, Confidence, Experience)** model, controlling DevOps projects in enterprise business environments.

The following chapters will be covered under this section:

- *Chapter 1, Defining the Reference Architecture for Enterprise DevOps*
- *Chapter 2, Managing DevOps from Architecture*
- *Chapter 3, Architecting for DevOps Quality*
- *Chapter 4, Scaling DevOps*
- *Chapter 5, Architecting Next-Level DevOps with SRE*

1
Defining the Reference Architecture for Enterprise DevOps

This chapter is an introduction to **DevOps architecture** for the **enterprise**. First, we'll look at the business of an enterprise. The business sets its goals and with that, defines the criteria for IT delivery, which supports these business goals. Therefore, the DevOps architecture must be aligned with the enterprise architecture. In this chapter, we will learn how to set up the reference architecture and design the different DevOps components while working with the **VOICE model**. Next, we'll learn how to deal with service levels and key performance indicators in DevOps models.

By the end of this chapter, you will have a clear view of how to start using the architecture and defining a DevOps strategy. An important lesson you'll learn in this chapter is that setting up DevOps in an enterprise becomes more complicated when organizations have outsourced large parts of their IT delivery. During this chapter, you will learn how to engage DevOps in enterprises with sourcing models.

We're going to cover the following main topics:

- Introducing DevOps in IT delivery
- Creating a reference architecture
- Introducing DevOps components
- Understanding SLAs and KPIs in DevOps
- Working with the VOICE model

Introducing DevOps in IT delivery

This book will focus on implementing and scaling DevOps in large enterprises. Before we get into the specific challenges of an enterprise, we need to have a common understanding of DevOps.

Somewhere, businesses and their leaders must have thought that it was a good idea to put developers and operators into one team. In essence, DevOps is the *development* and *operations* stages working as one team, on the same product and managing it. You build it, you run it.

DevOps has gained a lot of momentum over the past decade, especially in enterprises. But implementing DevOps turned out to be quite difficult. The reason for this is that enterprises are not organized in a structure that works for DevOps. From the last century onward, most enterprises outsourced a lot of their IT. Most of the IT muscles of a major enterprise are therefore still with system integrators and software houses. DevOps becomes more difficult when development is done by a software house and operations is outsourced to a system integrator.

DevOps starts with the business. By bringing teams together into a development and operations environment that traditionally work in silos, an enterprise can speed up development and release new products and services. The rationale behind this is that less time is needed to do handovers between development and operations. Also, by removing the barrier between development and operations, the quality of products will improve since DevOps includes *quality assurance*, *testing*, and *security*. Customer feedback is continuously evaluated and included in new iterations of the product.

The benefits of DevOps are as follows:

- It brings business, development, and operations together, without silos.
- Enterprises can respond faster to demands from the market because they're absorbing continuous feedback.

- Products are continuously improved and upgraded with new features, instead of planning for major next releases.

- Through automation in DevOps pipelines, enterprises can reduce costs in terms of both development and operations and, at the same time, improve the quality of their products.

It starts with the business and thus the starting point is the **enterprise architecture**. This is where the business goals are set and we define how these goals will be met. IT delivery is key to meeting these goals. In large enterprises, the architecture also defines the IT delivery processes and the demarcation between these processes. We will look at IT delivery and its processes in more detail in the next section.

Understanding IT delivery in enterprises

As we mentioned at the beginning of this section, large enterprises typically have an operating model that is based on outsourcing. This makes implementing DevOps more complicated. The enterprise architect will have to have a very clear view of the demarcation between the different processes and who's responsible for fulfilling these processes. Who is responsible for what, when, and why? The next question is, how does it map to DevOps?

First, we need to understand what the main processes are in IT delivery. These processes are as follows:

- **Business demand**: A business needs to understand what the requirements are for a product that it delivers. These requirements are set by the people who will use the product. Customers will demand a product that meets a specific functionality and quality. The architecture must focus on delivering an end product that satisfies the needs of the customers of an enterprise. IT delivery is a crucial part of delivering an end-product. In DevOps, an assigned product owner makes sure that the product meets the requirements. The product owner will have to work closely with the enterprise architect. In the *Creating a reference architecture* section, we will learn that the enterprise architecture and DevOps are complementary.

- **Business planning**: Once the demand is clear, the product needs to be scoped. In DevOps, product teams typically start with a **Minimum Viable Product (MVP)**, a first iteration of the product that does meet the requirements of the customer. When designing the MVP, processes need to be able to support the development and operations of that product. Hence, business planning also involves quality management and testing, two major components of IT delivery. This needs to be reflected in the architecture.

- **Development**: In DevOps, the product team will work with user stories. A team must break down the product into components that can be defined as *deliverables*. For this, we must have a clear definition of the *user story*. A user story always has the same format: *As a [function of the user] I want to [desire of the user] so that I [description of the benefits a user will get if the function has been delivered and the goal is achieved]*. The key of any user story is its acceptance criteria, or the **Definition of Done (DoD)**. When is the product really finished and does it meet the goals that have been set? In *Chapter 3, Architecting for DevOps Quality*, you will learn more about the DoD.

One important remark that must be made is that when we refer to a product, we are talking about a product that is code-based.

> **Tip**
> There's one major movement in IT delivery: everything in IT is shifting to code. It's one of the main principles of *The Modern DevOps Manifesto*: *Everything is code*. It applies to applications, but also to infrastructure components such as network devices, servers, and storage devices. Therefore, DevOps not only includes software development and operations for applications, but also for infrastructure with **Infrastructure as Code** and **Configuration as Code**. Public clouds such as AWS, Azure, and Google Cloud Platform play a significant role in these developments.

In other words, the team is developing code: application code, Infrastructure as Code, and also test code. A developer will work on a specific piece of code that has been defined in the product backlog. The whole end product – for instance, an application – has been broken down into **product backlog items (PBIs)**, where each developer will work on a PBI. As soon as a piece of code is ready, it needs to be tested on itself, but also as a component of the end product. Due to this, in development, code needs to be merged. This merging process is triggered by a *pull request*, where the developer requests to have the code merged and joined to the end product, thus fulfilling the user story. This is done using **pipelines**.

In *Chapter 2, Managing DevOps from Architecture*, we will discuss setting up and managing pipelines, both for application development and for infrastructure.

We can divide the whole DevOps cycle into two major phases called deployment and operations, as follows:

- **Deployment**: In this stage, the code is tested and validated so that it matches the user story. It will now be deployed to the production state. Testing and releasing to production is a process that, ideally, is automated in the pipeline, as is integration. Before the code is actually pushed to production, it also needs to be merged with configuration. Think of security packages that need to be applied to components that run in production. In the test and quality process, the full package – *application code* and *infrastructure components* – needs to be validated as "ready for production". The result should be a "live" product. If, when you're performing testing and validation, bugs, flaws, or violations of security rules are discovered, the product will be sent back to an earlier stage in the development process.

- **Operations**: After deployment, the live product needs to be operated on. For this, enterprises work according to **IT Service Management (ITSM)** principles. The fact that operators are in the same team as developers doesn't mean that the ITSM processes are not valid anymore. An example is when incidents occur and the incident management process must be triggered. In operations, we distinguish between the following main processes:

  a) Request fulfillment

  b) Incident management

  c) Problem management (postmortem)

  d) Configuration management

  e) Change management

But DevOps adds something to this; that is, **continuous integration and continuous delivery (CI/CD)**:

- **Continuous integration (CI)**: CI is built on the principle of a shared repository, where code is frequently updated and shared across teams that work in the cloud environments. CI allows developers to work together on the same code at the same time. The changes in the code are directly integrated and ready to be fully tested in different test environments.

- **Continuous delivery (CD)**: This is the automated transfer of software to test environments. The ultimate goal of CD is to bring software to production in a fully automated way. Various tests are performed automatically. After deployment, developers immediately receive feedback on the functionality of their code.

CI/CD requires a *feedback loop* to make it continuous. It needs feedback about the delivered products and services. This is then looped back to the developers and from there, new iterations are planned to improve the product or service.

This works well if an enterprise controls the full cycle, but large enterprises have outsourced a vast number of activities to other companies. The rationale behind this sourcing strategy is typically because a certain activity is not perceived as a core activity, and it can be done more cost-effectively by a company that specializes in such activities.

However, enterprises have gone through a massive change over the last decade. IT has become more and more important and, in some cases, has become a core activity. Banks are a good example. Banks are IT companies nowadays, and the output of their IT delivery is financial products. Due to customer demands, releases of these products with new features have become more frequent, with up to several releases per day. The consequence of this is a major shift in IT delivery itself.

The next few sections will discuss how IT delivery works in sourcing models and how it impacts successfully implementing DevOps.

IT delivery in sourcing models

In this section, we will look at the **sourcing model** in large enterprises. This can be quite complicated, but if we learn to think in terms of sourcing **tiers**, it becomes more tangible and comprehensible. This is the **target enterprise model**, as shown in the following diagram:

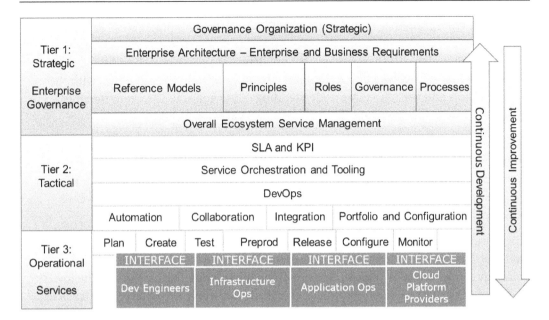

Figure 1.1 – Target enterprise model

Using this model, we can break down IT delivery into three tiers:

- **Tier 1**: **Strategic** level. This is the tier for enterprise governance. The enterprise defines the strategic business goals that are translated in the enterprise architecture. The overall architecture principles are the outcome of the enterprise architecture and drive the IT architecture, including DevOps. We will discuss this further in the *Creating the reference architecture* section.

- **Tier 2**: **Tactical** level. This the tier where the IT architecture is developed, including DevOps. It's also the tier where **service-level agreements** (**SLAs**) and **key performance indicators** (**KPIs**) are defined to measure the outcomes of IT delivery. You will learn more about this in the *Understanding SLAs and KPIs in DevOps* section.

- **Tier 3**: **Operational** or **services** level. At this level, the components of the architecture are detailed, including the interfaces to the various suppliers and service providers. The agreements that are defined in tier 2 must be adopted at this level so that all involved developers and operators work in the same way, with the same tools and with the same understanding of the goals. In the *Understanding DevOps components* section, we will learn more about this.

In practice, we see service providers also acting on tier 2; for instance, if they are involved in larger programs spanning multiple products. Tier 2 then becomes the orchestration level, where a provider is responsible for aligning different streams in the lower tier. The key takeaway is that tier 1 should always be the enterprise level, where the overall governance and architecture is defined.

In this section, we learned that a lot of enterprises have outsourced larger parts of their IT and that this can complicate the process of implementing DevOps. We learned that the strategy for the entire enterprise is at tier 1, the highest tier. DevOps sits on the lowest tiers, where projects are actually executed. This is the tier where the enterprise interfaces with sourcing companies. However, this only works if we have a clear view of the architecture. We will discuss this in the next section.

Creating a reference architecture

In the previous sections, we looked at the different processes in IT delivery, how it is integrated with DevOps, and how this is executed in sourcing models. We have learned that it starts with a clear architecture that clearly defines the processes.

Any **DevOps architecture** will have to address planning, development, integration, deployment, and operations. But we have to keep in mind why we are doing DevOps, which is to achieve business goals in a faster, more agile way where we continuously improve products. The DevOps architecture does not stand on its own; it has to be linked to the enterprise architecture.

An enterprise architect will most likely start from **The Open Group Architecture Framework** (**TOGAF**). TOGAF is globally accepted as the standard for enterprise and business architecture. It uses the **Architecture Development Method** (**ADM**) to draft the architecture. The ADM is shown in the following diagram:

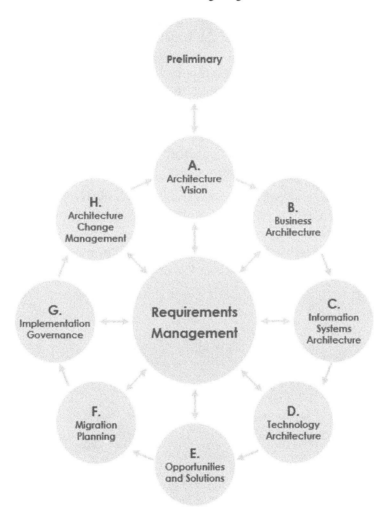

Figure 1.2 – The ADM cycle in TOGAF

Just like DevOps, the ADM is a cycle – except for the preliminary phase, which is where the need for an architecture is formulated. This has to be done at the tier 1 level in the sourcing model that we discussed in the previous section, *IT delivery in sourcing models*. The strategy and enterprise architecture is always set at tier 1.

The core of ADM is the sequence in *architecture design*, which is business first, and setting the principles and requirements for the actual solutions. These principles drive the architecture for data usage, the applications, and finally the technology that will be used. This is important because architecture is not about technology in the first place. Technology is purely the ability to achieve business goals at the enterprise level.

ADM assumes that the architecture is not static. It changes as soon as the business requirements change. If the business demands change, there will likely be a need to adapt the architecture and the forthcoming solutions. This is where TOGAF and DevOps meet, since the two frameworks complement each other.

The following table shows where the enterprise architecture and DevOps are complementary. To put it very simply, the enterprise architecture sets the *strategic business goals*, where DevOps translates this at a more tangible, tactical level and really tells us how to achieve these goals by developing products and executing operations. The following table shows the main differences between the **enterprise architecture (EA)** and DevOps:

Enterprise Architecture	DevOps
EA translates the business goals into the enterprise IT architecture. The EA is valid for the entire enterprise and its business as a whole.	DevOps is organized and executed at the level of the business unit.
EA starts at a very high level, capturing the business goals, and then drills down to the applications and, at the very end, the technology. For that, EA defines different projects.	DevOps starts with the projects defined in the EA and sets these as backlog items that can be executed by one DevOps team.
Change management in EA involves changes that impact the business of the entire enterprise.	Change management in DevOps involves making changes to the product, as delivered by the DevOps team, specifically for one business unit.
Changes to the EA occur frequently. But since they impact the business as a whole, the EA will have to assess changes carefully, The enterprise architect will have to involve all the stakeholders of the enterprise.	In DevOps, changes are actively sought after as part of the continuous feedback loop and will drive the continuous improvement of the product and the way it is delivered.
EA sets architecture principles for the whole enterprise.	DevOps will need to adhere to the EA architecture principles, but since DevOps teams are expected to be self-directing, they can use other principles that are complementary to the EA. Yet, these principles may not conflict with EA.
EA needs to have values that can be measured at enterprise levels.	The values of DevOps teams are measured by product and at the business unit level, but have to contribute to the enterprise as a whole.

In the next section, we will study the DevOps principles.

Understanding the DevOps principles

The enterprise architecture is executed on tier 1, the strategic level. This is where the goals are set for the entire enterprise. The next level is tier 2, where DevOps teams will translate the goals into product features and start developing. DevOps teams will have to work according to a set of principles.

In this section, we will look at the main principles for DevOps. In this book, we will use the six principles from the **DevOps Agile Skills Association (DASA)**:

- **Customer-centric action**: Develop an application with the customer in mind – what do they need and what does the customer expect in terms of functionality? This is also the goal of another concept, domain-driven design, which contains good practices for designing.

- **Create with the end-result in mind**: How will the product look when it is completely finished?

- **End-to-end responsibility**: Teams need to be motivated and enabled to take responsibility from the start to the finish of the product life cycle. This results in mottos such as *you build it, you run it,* and *you break it, you fix it.* One more to add is *you destroy it, you rebuild it better.*

- **Cross-functional autonomous teams**: Teams need to be able to make decisions themselves in the development process.

- **Continuous improvement**: This must be the goal – to constantly improve the product.

- **Automate as much as possible**: The only way to really gain speed in delivery and deployment is by automating as much as possible. Automation also limits the occurrence of failures, such as misconfigurations.

Adhering to these principles will lead to the following architecture statements, which are at the core of DevOps:

- **Automation**: Following the principle of "everything is code," the next step is "automate everything." With automation, the amount of time between testing and deployment will be significantly reduced, enabling a faster release process. But automation will also lead to less manual interaction and therefore less errors.

- **Collaboration**: Two of the six principles are cross-functional autonomous teams and end-to-end responsibility. This can only be achieved by collaboration. Development and operations will have to work very closely together to speed up the delivery of releases. Although this is also a cultural change, collaboration requires a common toolset that supports collaboration.

- **Integration**: Development and operations come together, but also, business and IT come together. In DevOps, we integrate the business demands with IT delivery using *user stories*. Code is integrated with new functionality that is coming out of business demand. That demand is changing faster these days, so development needs to keep up by means of CI. This will lead to changes in operations as well – they will need to adopt these new developments at the same speed. With that, integration is the core of DevOps.

- **Portfolio and configuration management**: Automation and integration require a clear portfolio that contains the building blocks that can be automated easily. These building blocks are artifacts in the ADM cycle, and they represent packages of functionality that can be used to fulfill a requirement. A building block is reusable and replaceable; therefore, it must be clearly and specifically defined. Better said, the configuration of the building blocks needs to be well documented and brought under the control of configuration management. If done well, these building blocks will also have clear interfaces so that they can be fully automated.

In this section, we looked at the IT delivery processes and how they impact DevOps. We learned that IT delivery is driven by business demand and that this business demand is the starting point for any architecture. This is included in the TOGAF framework for the enterprise architecture. After that, we mapped the enterprise architecture to DevOps principles.

In the next section, we will merge the DevOps principles for the architecture and the IT delivery principles into a reusable reference model.

Working with the DevOps architecture reference model

The final step is to merge the DevOps principles into one model for our **reference architecture**. The model contains two circles. The outer circle is the *product circle*, while the inner circle represents the *operational activities*. As a logical consequence, the outer circle is governed by the enterprise itself.

The inner circle is about actually delivering the products using DevOps. There are interfaces between the outer and inner circle: collaboration, automation, integration, and configuration management.

The reference model is shown in the following diagram:

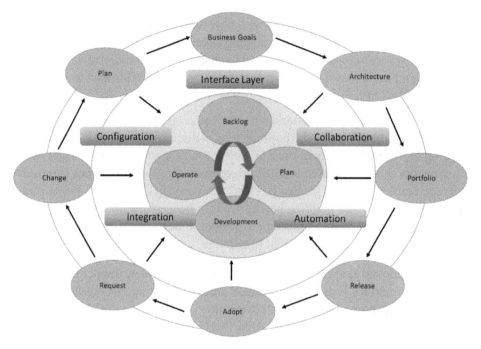

Figure 1.3 – The DevOps architecture reference model

In the outer circle, the business goals are translated into the architecture. From the architecture, a portfolio is created with building blocks to create products and services. Products are released to the market and adopted, but due to changing demands, there will be requests for changes. These changes will drive enterprise planning and ultimately change the business goals, meaning that the business will constantly have to adapt to changing demands. This is the field of enterprise architecture.

The plans and the actual builds are executed in the inner circle. In this circle, the product is broken down into product backlog items that will be developed and eventually operated on by DevOps teams. These teams do not operate by themselves, but on *triggers* from the outer circle. That's what the interface layer is about – it's the interface between the business and the execution teams doing IT delivery. There's collaboration between architecture and development. Releases should be automated as much as possible, requests and changes must be integrated with planning and the backlog of the DevOps teams, and builds that are pushed to production must be monitored and brought under the control of configuration management so that the architecture and portfolio stay consistent in case of changes.

Let's have a look at how this would work in practice by plotting personas into the model. This will result in a *DevOps workflow* for enterprises, as shown in the following diagram:

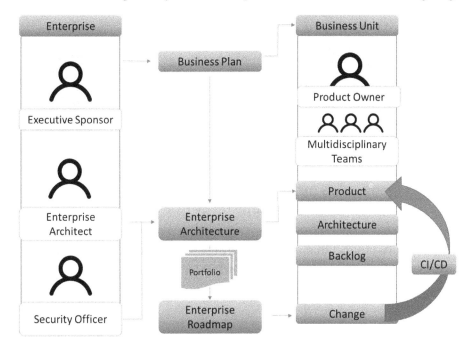

Figure 1.4 – DevOps workflow for enterprises

Here, we have created a model where the enterprise has full control over its portfolio and products. Yet, it can improve quality and speed up delivery by working with combined, **multidisciplinary teams** – even those that come from different suppliers.

In the next section, we will study the final, lowest tier in our model and discover various DevOps components.

Introducing DevOps components

So far, we've learned how to start defining the architecture, looked at the architecture principles for DevOps, and drafted a reference architecture model. The next step is to look at the different **components** within DevOps. In this section, we will learn what components must be included in a DevOps architecture. This is tier 3 of our target enterprise model – the level where all the activities are executed.

The following diagram shows all the components that will be discussed briefly:

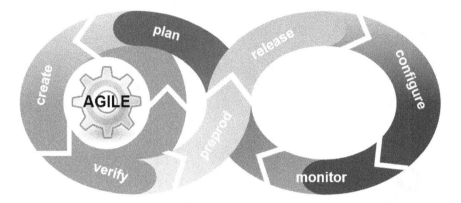

Figure 1.5 – The DevOps life cycle

The reason that this has been presented as an infinite loop – or a *pretzel* – is because feedback from the live product that is managed by *ops* (operations) will be continuously looped back to *dev* (development) to improve the product.

The different components are as follows:

- **Plan**
- **Create** (in some DevOps models for components, this is referred to as **Code and Build**)
- **Test** (in some models, this is referred to as **Verify** or **Validate**)
- **Preprod** (in some models, this is referred to as **Pre-release**)
- **Release**
- **Configure**
- **Monitor**

At this level, interoperability is crucial. Remember that large enterprises will likely work with several service providers, fulfilling parts of the IT delivery process. When we want all these companies to work together in a DevOps way, we need to make sure that the processes and tools are aligned. Next, we need to have a common understanding of the various activities that are executed as part of these processes. The key term here is *consistency*. All DevOps components must be defined and implemented in a consistent way. Every developer and operator must work according to the same definition and with the same components.

The main question is, in what stage should ops already be involved? The answer is, at the *earliest stage possible*, so indeed in the plan phase. Ops plays a key role in defining how products can be managed once they've gone live. They should set requirements and an acceptance criterion before going live. If a developer builds something that can't be managed by ops, the product will fail, and business demands will not be met.

The following table breaks down the components into activities:

Component	Activities
Plan	
	Define production metrics, including the definition of SLAs and KPIs.
	Define requirements (use case, prototyping).
	Evaluate business metrics (examples include resolution times and customer satisfaction).
	Update release metrics.
	Prioritize new features, functions, and fixes.
	Plan release.
Create (Code and Build)	
	Design software and configuration, including Infrastructure as Code.
	Code, code merge, code quality, and performance.
	Build.
	Perform functional tests.
	Release to test procedure.
Test	
	Acceptance test.
	Regression test.
	Quality analysis.
	Security analysis.
	Performance test.
	Defect status.
	Configuration test.
	Release test.
PreProd	

		Release to staging, using auto-release where the release is automatically pushed or a triggered release when the requirements of the "ready state" are met.
		Release for approval and push to production.
Release		
		Schedule the release.
		Deploy application code.
		Check deployment status.
		Set change controls.
		Validate the fallback or recovery plan.
Configure		
		Define and apply infrastructure provisioning and configuration, including storage, database, and network.
		Define and apply application provisioning and configuration.
Monitor		
		Monitor the performance and availability of the IT infrastructure, network, and application.
		Monitor the end user's response and experience.

In *Chapter 2, Managing DevOps from Architecture*, and *Chapter 3, Architecting for DevOps Quality*, we will dive deeper into this and how architects can improve their designs for these components using CI/CD pipelines to enable automation, collaboration, and integration.

In the next section, we will discuss the drivers for architecture from a business perspective, as laid down in SLAs and KPIs.

Understanding SLAs and KPIs in DevOps

In the *Understanding IT delivery in enterprises* section, we learned that in DevOps, IT delivery and IT service management processes are still valid. Typically, enterprises contract SLAs and KPIs to fulfill these processes so that these enable the business goals. If one of the processes fails, the delivery of the product will be impacted and as an ultimate consequence, the business will not achieve its goals, such as an agreed delivery date or *go live* release of the product. Hence, understanding SLAs and KPIs is important for any architect. This is why it is included in the sourcing model that we discussed in the *IT delivery in sourcing models* section.

Service-level agreements are positioned between the tactical processes of DevOps and the strategic level at the enterprise level where the goals are set. SLAs and KPIs should support these goals and guide the DevOps process.

The six most important metrics that should be included in SLAs for DevOps are as follows:

- **Frequency of deployments**: Typically, DevOps teams work in sprints, a short period of time in which the team works on a number of backlog items as part of the next release of a product. The KPI measures how often new features are launched on a regular basis. Keep in mind that releases of new features can be scheduled on a monthly (often spanning multiple sprints), weekly, or even daily basis.

- **Deployment time**: The time that elapses between the code being released after the test phase to preproduction and ultimately production, including the *ready state* of the infrastructure.

- **Deployment failure rate**: This refers to the rate of outages that occur after a deployment. Ideally, this should be zero, but this is not very realistic. Deployments – especially when the change rate is high – will fail every now and then. Obviously, the number should be as low as possible.

- **Deployment failure detection time**: This KPI strongly relates to the previous one. Failures will occur, but then the question is, how fast are these detected and when will mitigating actions to resolve these issues be taken? This KPI is often also referred to as **Mean Time to Recovery (MTTR)**. This is the most important KPI in DevOps cycles.

- **Change lead time**: This is the time that elapses between the last release and the next change to occur. Subsequently, it is measured in terms of how long the team will need to address the change. Shorter lead times indicate that the team works efficiently.

- **Full cycle time**: The total time that elapses for each iteration or each deployment.

This list is by no means exhaustive. Enterprises can think of a lot of different metrics and KPIs. But the advice here is to keep things simple. Keep in mind that every metric that is included in any contract needs to be monitored and reported, which can become very cumbersome. One more thing to remember is that the most important metric sits at the business level. Ultimately, the only thing that really counts is how satisfied the customer of the business is or, better said: *what's the value that's delivered to the end customer*?

In the final section of this chapter, we will elaborate on the term *value* by explaining the **VOICE model**.

Working with the VOICE model

DevOps teams need to deliver value to the end customer. The VOICE model, as defined by the IT company *Sogeti*, addresses this. VOICE stands for **Value, Objectives, Indicators, Confidence**, and **Experience**. The idea behind this model is that any IT delivery should deliver value to someone – typically, the end customer of a business. Value sets the objectives for IT delivery and these objectives are measured using indicators. Indicators also measure whether the pursued value will be achieved.

Confidence is about the indicators and if they contain relevant information to confirm that IT delivery actually results in the targeted value. Lastly, experience tells us if the delivered system is fulfilling the business demands and which improvements will lead to more business value. With that, the cycle starts over again.

This model is shown in the following diagram:

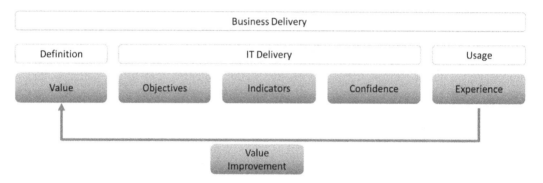

Figure 1.6 – The VOICE model

Since VOICE also involves looping feedback back to the beginning of the cycle with the aim of improving products and adding more value to the business, the model can be used for DevOps projects.

In *Chapter 3, Architecting for DevOps Quality*, we will explore VOICE in more detail.

Summary

This chapter was the introduction to DevOps for architects. We learned that the enterprise architecture sets the architecture principles for the entire enterprise by using the TOGAF methodology. The business goals are defined at the enterprise level. DevOps projects and teams are concerned with IT delivery and fulfilling the business' demands by building, deploying, and running IT systems.

DevOps needs to adhere to the business goals and, therefore, with the enterprise architecture. Yet, DevOps features a specific architecture that enables CI/CD in IT systems. Because of this, we learned about the six DevOps principles and how these are applied to a reference model in which the enterprise still has full control of the products, but multidisciplinary teams can work autonomously on them.

Next, we looked at the different DevOps components and KPIs to measure the outcomes of DevOps projects. The key takeaway from this is that every project needs to add to a better user experience and thus add business value. Due to this, we briefly studied the VOICE model.

In the next chapter, we will learn more about automation, collaboration, and integration by designing CI/CD pipelines.

Questions

1. *True or false*: DevOps brings business, development, and operations together, without silos.

2. The DevOps principles lead to four key attributes in the architecture for DevOps. One of them is automation. *Name the other three.*

3. What does the acronym *CI/CD* stand for?

4. Ideally, every deployment succeeds, but in practice, some deployments will fail. The time that is needed to detect this failure and start mitigating actions is an important KPI in DevOps contracts. *What is the commonly used term for this KPI?*

Further reading

*   *The Modern DevOps Manifesto*: https://medium.com/ibm-garage/the-modern-devops-manifesto-f06c82964722

*   *Quality for DevOps Teams*, by Rik Marselis, Berend van Veenendaal, Dennis Geurts and Wouter Ruigrok, Sogeti Nederland BV.

2
Managing DevOps from Architecture

In the previous chapter, we learned about the different DevOps components, which comprise automation, collaboration, integration, and configuration management components. In this chapter, we will learn in more detail how to design these components and how to manage the **DevOps cycle** from these components. We will learn that automation and integration start with standardizing building blocks, called **artifacts**. These artifacts are linked with a portfolio that is defined by the **enterprise architecture**. Before we get to launch **DevOps projects** using automation and integration, we need to understand the business strategy and demand for architecture.

After completing this chapter, you will be able to identify the different components of demand management and how this drives portfolio management. You will also learn what the various stages are in **continuous integration** (**CI**) and how automation can help enterprises in speeding up deployment processes. In the last section, you will see that collaboration in these processes is crucial, and we will look at how you can enable teams in working together, introducing (as an example) **Kanban** and other frameworks.

In this chapter, we're going to cover the following main topics:

- Assessing demand as input for the architecture
- Designing and managing automation

- Implementing and managing configuration management

- Designing and managing integration

- Designing and managing collaboration

Assessing demand as input for the architecture

You can't just start with a DevOps project—a business will need to know what they want to achieve before they launch projects. For that matter, there's no difference between a traditional Waterfall project and DevOps in an *Agile* way of working—you need to know where you're going. That's a very simple explanation of something that is called **demand management**. In this section, we will learn about demand as input for an architecture and how this leads to projects.

Demand management can be defined as a process wherein an enterprise collects and prioritizes ideas to improve business outcomes. To be able to do that, the enterprise needs to assess the demands from the *outside*, meaning the market—in other words: *What do customers want?* But it also needs to assess whether the current portfolio is still up to date and that ongoing projects will still deliver the desired outcome. **Portfolio management** is a constant evaluation of market demands and ongoing activities. The modern challenge is that this evaluation has to be done at a higher speed than, let's say, a decade ago. Demands are changing fast and that does impact portfolio management drastically.

Portfolio management includes at least the following components and processes (Romano, L., Grimaldi, R., & Colasuonno, F. S. (2016). *Demand management as a critical success factor in portfolio management*. Paper presented at PMI® Global Congress 2016—EMEA, Barcelona, Spain. Newtown Square, PA: Project Management Institute):

- Definition of **enterprise vision** and **strategy**: This is part of the enterprise architecture and defines the strategy for the enterprise.

- **Demand management**: The processes to collect ideas and identify opportunities for future products and services that are in the portfolio.

- **Ongoing components**: Validation of whether existing products and services are still relevant for the enterprise and its customers.

- **Components assessment**: This concerns the business case. What will be the investment in developing new products and services or changing existing components, and what is the expected revenue after the release or change?

- **Budgeting**: The calculation of the required amount of resources needed to start development or realize upgrades of existing components in the portfolio.

- **Prioritization and selection**: Development and changes need to be prioritized against the strategy of the enterprise. What is a must-do and what is a nice-to-have, and in what timeframe?

- **Portfolio governance and communication**: This relates to who's responsible for what in demand and portfolio management. A recommended approach is to have a **Responsible, Accountable, Consulted, and Informed (RACI)** matrix for portfolio management: Who's responsible for which component? Who is accountable? Who should be consulted? Who should be informed?

- **Portfolio implementation**: Adding the new or changed components to the portfolio in a structured acceptance process. Are the components documented in an appropriate way? Are the components signed off by the responsible owner? Have all acceptance criteria been met?

- **Portfolio reporting and review**: The reports should reflect whether the portfolio is aligned with the business strategy. Is the portfolio still relevant to the strategy and the desired business outcomes? Or, does the enterprise need to change components, or maybe even consider adapting the whole portfolio (for instance, by divesting parts of the portfolio)?

- **Benefits realization**: What has been the real value of the portfolio and the benefits to the enterprise and its business?

The portfolio drives the projects in an enterprise, and also DevOps. The key to DevOps is automation so that projects can keep up with the speed of business changes and demands to **Information Technology (IT)**. Crucial in automation is the creation of standardized building blocks. From the portfolio, we need to define these building blocks, called artifacts. **Configuration management** is about managing these artifacts so that they fit to the portfolio. To manage this, we need version control. We will discuss this in the *Implementing and managing configuration management* section.

We now have the cycle of creating a portfolio to manage configuration items and enable automation. This is shown in the following screenshot:

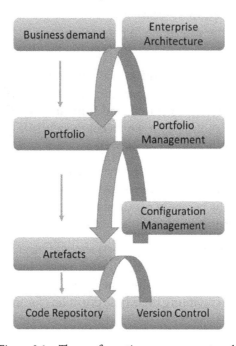

Figure 2.1 – The configuration management cycle

The **code repository** holds the standard building blocks that are under version control. This is a requirement for enabling automation. In the next section, we will learn more about automation.

Designing and managing automation

In this section, we will discuss **automation** for DevOps. For starters, automation is not only about tools, although we will discuss tools at the end of this section. The first questions that architects will need to answer regard *what* they need to automate and *why*. It's not about the tools but about the process.

First, we need to answer the following question: *Why would we need automation?* The answer to that question is because of *standardization*. The reason for businesses to adapt DevOps is because they want to speed up delivery processes. The only way to do that is by standardization of building blocks, workflows, processes, and technologies. By implementing and adhering consistently to standards, companies will limit varieties in the delivery chain and can then start automating it. The big trick in automation is cutting down the waiting time.

Before companies turned to DevOps, IT delivery was driven by waiting time. Developers had to wait for a server. In a worst-case scenario, the server needed to be ordered first, then installed and configured before it could be released to the development team. Then, developers could do their work and deliver the software to testers, but then had to wait for the testers to come back with results. Finally, the product was ready for release, but now the team would have to wait for the final *go/no-go* decision.

In DevOps and automated **CI/CD pipelines** (where **CD** is an acronym for **Continuous Delivery**), the waiting time is strongly reduced by standardization, if done well. Architects will have to keep in mind that a pipeline consists of two major components: *software* and *infrastructure*, even though infrastructure has become code too. Typically, we will work in a cloud where we're not deploying physical servers, but **Infrastructure as Code** (**IaC**).

Everything has become code. Automation of infrastructure, configuration, and deployment of software is the core of DevOps. Only by using automation can a company speed up delivery to periods of weeks, days, or even hours.

The following screenshot shows the automation process for the CI/CD pipeline. There are two major components in this pipeline: the **deployment pipeline** and the **infrastructure pipeline**:

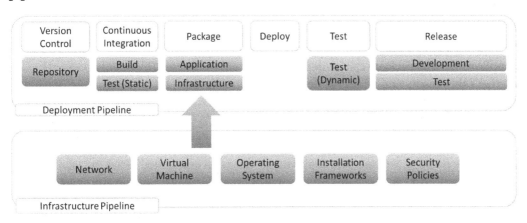

Figure 2.2 – Deployment pipeline and infrastructure pipeline

The software is developed, tested, and deployed in the deployment pipeline. That software needs to land on infrastructure—for example, on a **virtual machine** (**VM**) in a public cloud such as **Amazon Web Services** (**AWS**) or Azure. At a certain point, we will have to merge the software code with the infrastructure and configuration packages for that infrastructure. That whole package is tested, validated, and eventually pushed to the production stage.

> **Note**
>
> *Figure 2.2* shows how important testing is. There are a couple of touchpoints that include testing procedures. In the *Designing and managing integration* section, we will explain more about the various tests in CI, such as static and dynamic analysis. *Chapter 3, Architecting for DevOps Quality* also has a section about executing tests.

So, the deployment pipeline is used to convert code to a deployable application package, deploying the package, validating the package, and releasing the package to production.

In the infrastructure pipeline, we provision the environments on which the application package can be deployed. In *Figure 2.2*, VMs are mentioned as an example, but infrastructure can also consist of **containers** or **serverless components**. In fact, native applications will use containers and serverless environments over VMs.

Understanding pipeline components

Let's take a better look at the diagram shown in *Figure 2.2* and comment on the different components. The pipeline starts with **version control** and **configuration items**. In the next section, *Implementing and managing configuration management*, we will discuss this in more detail.

The actual deployment starts with the package containing all the files that we need to run the application. This package can be deployed to an environment. *To any environment?* Well, that's the reason to abstract the deployment pipeline from the infrastructure. Ideally, we would like to be able to run code on different platforms, so we need to code in such a way that it can run on infrastructure in—for instance—AWS, Azure, or on a private stack, or even using different operating systems such as Windows or Linux distributions.

If we have the package, we can start deployment: this is an automated series of tasks to deploy our application on a test or staging environment. Here, we can run automated simple tests—often referred to as smoke tests—to verify that the code is actually running. This simple test will not be sufficient to validate all the required functions and to detect bugs: this is done in the test phase. We will learn more about testing in the *Designing and managing integration* section.

If all tests have been successfully executed, we have a validated application package that is ready to be pushed to the production stage.

The following aspects need to be considered in terms of automation:

- **Code development**: Automation in DevOps starts as soon as developers begin developing the code. The code needs to be "automation ready," meaning that from the moment the code is checked in, the actual build is triggered and runs through automated tests and code validation. Next, the code is compiled and brought under version control. Ultimately, when all tests have been executed and the full package is validated, it will be pushed to production. Developers need to take this automation sequence into account when they develop the code. After the code is deployed into production, it needs to be monitored. So, as soon as the package is "prepared" for go-live, scripts will be merged to the package to enable it to be monitored, and for logs to be collected.

- **Continuous testing**: In DevOps cycles, software will constantly be evaluated and improved. That's what DevOps is about, after all: increasing agility and CD. Code will therefore be changed regularly. Every time code is changed, it needs to be tested and validated. That's where continuous testing comes in. Automated testing is used to track and predict issues in code changes, run multiple tests, and ultimately release approved automated builds.

- **Monitoring**: It's a misunderstanding that monitoring is only needed when applications are pushed to production. Monitoring is relevant during the whole CI/CD life cycle and is a crucial component in automation. Monitoring is required to track events, identify causes for why code is malfunctioning, prioritize events, and proactively suggest actionable improvements.

In short, by applying automation, we can bring release timelines down from months or weeks to just a few days, or even hours. This is only possible if we automate as much as we can. But automation does more than just speed up delivery; it also reduces the risk of manual errors and will enable companies to achieve more consistency and higher reliability for the products they release. DevOps is not only about agility and speed but maybe even more about quality, by creating products with more accuracy in a repeatable process that can be delivered at a high pace.

Choosing the DevOps automation toolset

Now, so far, we haven't talked about **tooling**. The selection of the right tools is indeed very important. The challenge, however, is that there are so many tools available on the market. One of the tasks of an architect would be to guide the selection. The first decision that an architect will have to make is whether they want the automation to be "single-stack" or whether to opt for multiple tools for different automation areas.

Just look at the **periodic table** of **DevOps tools** (by *Digital.ai*) in the following screenshot. For each of the DevOps domains, there's a huge choice of toolsets. There are tools for managing the code, creating packages, automating the merging of pipelines, testing code, bug fixing, provisioning of infrastructure, managing infrastructure, and monitoring:

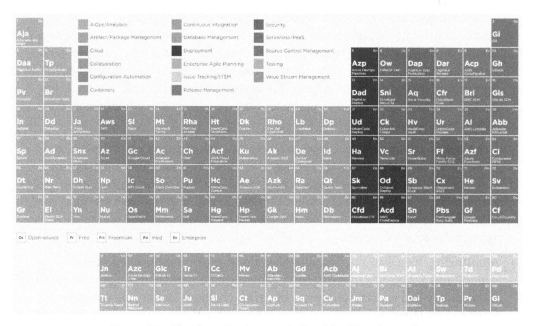

Figure 2.3 – The idea of having a periodic table of DevOps tools

Organizations will need tools to manage the DevOps process from end to end. There are tools that promise this end-to-end capability, but a set of tools is typically required, from the code build and commit to testing, deployment, and operations. One important thing to keep in mind is the level of integration between the tools. This depends not only on the source code that developers use, but also on the target platform. If the target platform is Azure, it makes perfect sense to use **Azure DevOps** and **Azure Resource Manager** (**ARM**) to deploy infrastructure on that platform. If the main platform is AWS, then tools such as Chef and Puppet, in combination with **AWS CodeDeploy**, are probably a better choice.

Then again, there's a major shift going on from using VMs to containers and serverless technology. A lot of DevOps projects use containers to deploy code, by using Docker and Kubernetes as the container *orchestration platform*. Containers are much more agnostic to the underlying infrastructure or cloud platforms than VMs, but also here, platforms all have their own Kubernetes engines. Think of **Azure Kubernetes Services** (**AKS**) on Azure and **Elastic Kubernetes Services** (**EKS**) on AWS.

One more important thing to consider is the type of tool that will be chosen. Open source tooling is very popular, but it's good to consider whether these tools meet the enterprise requirements. That differs per domain. Azure DevOps Pipelines and AWS CodePipeline are perceived as enterprise tools, but tools for version control (Git, GitHub, GitLab, **Subversion** (**SVN**), Cloud Foundry) are mainly open source. There can be good reasons for an enterprise to go for open source, as outlined here:

- **Community driven**: Large communities constantly improving the software. Enterprises can benefit from this.

- **Cost-effective**: There are typically licenses required for open source software, but looking at the total costs, open source is often a very good deal.

- **No risk of lock-in**: This is becoming more and more important for organizations. They don't want to be completely locked into the solutions of one software provider or the ecosystem of that provider. Open source allows an organization a great amount of freedom.

The negative side of open source that is often cited is that this software would be less secure and less stable than enterprise software. However, because the software is constantly reviewed and improved by a community, we find that open source software is just as secure and stable as non-open source software.

Implementing and managing configuration management

In the previous section, we learned that automation starts with version control and configuration items that form an application package in an artifact's repository. In this section, we will study how we can manage these artifacts.

Automation can only be done when **building blocks** (artifacts) and processes are standardized. **Standardization** requires three components, outlined as follows:

- **Portfolio** and **portfolio management**: A portfolio is the translation of the business strategy and the products that a business delivers to its customers. Those products consist of several artifacts: product components and processes. So, a portfolio is at the strategic level of an enterprise, whereas products and artifacts sit at a tactical level. A portfolio is defined by the enterprise architecture, products, and artifacts that are managed at a business-unit and project level. In short, products can't exist without a definition in the portfolio, but a portfolio is not something that we can automate in projects and pipelines. We can automate artifacts and even products, as long they are standardized and managed through version control.

- **Version control**: In the first section of this chapter, we learned that artifacts are derived from a portfolio. Artifacts need to be put under version control. A centralized version control repository is absolutely the number-one priority in CI/CD. This repository is a **single source of truth** (**SSOT**) for all configuration items that will be used for the development and building of the deployment packages. Creating packages with configuration items is done using the file versions in the **version control system** (**VCS**). Version control is mandatory to keep all configuration items consistent. A build produces a versioned package that can be retrieved from the repository.

- **Configuration items**: The term *configuration item* comes from the **IT Infrastructure Library** (**ITIL**), which is an IT library for IT service management. Configuration items are assets that need to be managed in order to deliver an IT service—basically, everything that is needed to execute a build and deliver an IT service. It includes building blocks for the application code, but also images for the operating systems, configurations for network settings, security packages, licenses, database settings, and test scripts. All these items need to be managed. Configuration items are defined by a **unique identifier** (**UUID**), the type of an item (hardware, software, script, license, and so on), a clear description, and the relationship that the item has with other items. To keep configuration items consistent, they need to be verified and validated at regular intervals or even in "real time." A best practice is to automate this, as we have already seen in the previous section. Agents will constantly monitor all assets in an environment and update the status of configuration items in the repository, commonly referred to as the **Configuration Management Database** (**CMDB**).

How is this linked to the enterprise architecture and the portfolio? Let's take security as an example. From the enterprise strategy, the security standards must be defined—for instance—to which industry framework the enterprise should be compliant with. That is translated into security policies that need to be set in the security packages. These packages containing security policies and rules are known as configuration items. In the third part of this book, about **DevSecOps**, we will learn more about this.

Let's make this a little bit more tangible. We have an application that runs on a VM and is attached to a database. The different configuration items could then be listed as follows:

- Application code
- VM template
- Operating system image
- Security settings for the application
- Security settings for the VM

- Access rules

- Database image

- Licenses for the operating system

- Licenses for database

- Network settings (**Internet Protocol** (**IP**) address allocation; network ports; routing)

This list is by no means exhaustive—these are just examples. All these items need to be managed to keep them consistent. The CMDB enables verification and validation of all items. By applying version control, we make sure that new builds only use verified configuration items—for example, a certain version of an operating system or a specific version of the application code.

A key takeaway here is that every configuration item is linked to the business strategy, portfolio, and enterprise architecture.

In this section, we've learned what configuration items are and how we manage these items from a single repository (CMDB) so that they remain consistent. In the next section, we will see why version control and consistency are so important.

Designing and managing integration

In this section, we will learn more about **CI**. First, we will look at the development and deployment of **application code**. Next, we will learn about the integration of **code pipelines** for applications and the infrastructure. Somewhere, the application code and the IaC have to merge together with specific configuration packages. Only then will we have a fully integrated model.

Let's look at a definition of integration first. This refers to an automated series of tasks to version, compile, package, and publish an application. This includes testing, whereby **unit tests** are used to validate that existing code performs well without interruptions. **Integration tests** run to ensure no integration issues occur. Additional checks, such as **static code analysis**, can be included as well to increase quality and feedback.

The CI/CD pipeline—and with that, the automation—starts with a merge of **source code**. This code is transformed to a build, as a product of an executable package that integrates with other executable packages. Code is pulled from the repository, analyzed, and committed to a build server, where a series of automated tasks will be executed to push to code into production, including the merge with the infrastructure.

As stated, testing and validation of the code is a crucial part of the build. Code is tested to verify that all risks are identified and mitigated so that the build meets enterprise compliance, defined in the enterprise architecture and, subsequently, the security policies. The next step is that the code is deployed to a development or test environment. As we have seen, this is preferably done in an automated way. The code is brought under version control and committed back to the repository.

Although it requires a lot of manual tasks, it's important to maintain a wiki—a collection of documents that are maintained by the developers—as part of the version control, containing the release notes with the build number, and issues that have been encountered and mitigated after the tests. The outcomes of the tests should be listed in the release notes.

To summarize, the CI process consists of the following steps:

1. **Detecting new committed code**: The source code in the application repository is forked, which means that the code is committed to the automated pipeline. This starts a sequence of build tasks. The next step is to validate the code. This is the **pull process**, whereby the code is checked out and placed on a **build environment**—in most cases, a **development server**. Static analysis is executed, which is a test without actually running the code. This step is also referred to as **linting**.

2. **Source code quality analysis**: The code is looked at to check it meets security and compliance requirements.. Part of this quality analysis is also a scan of the software used: is it licensed and is it compliant with the enterprise standards and software portfolio?

3. **Build**: This is the step where code is compiled and packages are assembled. The actual build begins with pulling the executable packages from the artifact's repository. This repository holds all the standards and policies that an enterprise uses to deploy code. The source code is merged with these packages. As an example, these packages contain the images for the operating systems, but also the security policies, such as hardening templates to protect systems from attacks. Packages for infrastructure are often stored in a repository that holds the images: the trusted image registry. These images are used to assemble an executable package to final deployment.

4. **Unit tests**: Unit tests ensure that the executable package runs without issues.

5. **Integration tests**: In the code validation, we executed a **static analysis**; now, a dynamic analysis is performed. The assembled package is tested to verify that it runs without errors, and test reports are generated.

6. **Creating a runtime executable file**: We're getting close to the final steps, which is the release of the code to the next phase. Before we can do that, runtime executables are generated so that the code can be started in a specific runtime order or routine. These runtime executables are often stored in a separate runtime library.

7. **Build notification**: The last step before deployment is notification that the build is ready, meaning the code and the executable packages have passed all tests. The package is ready for deployment. If the integration is stopped at any time during the automation sequence, developers will be notified so that they can fix any issues and then recommit the code. Every step, as described, is then repeated.

The integration process is shown in the following screenshot:

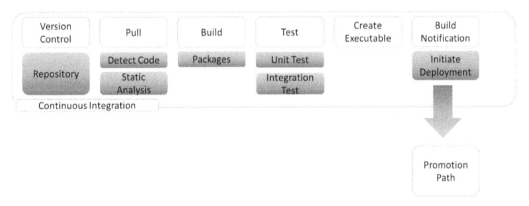

Figure 2.4 – The CI process

Figure 2.4 shows that the output of the CI path is directed toward something called a **promotion path**. Applications are seldom pushed directly to production environments. The application release is usually "staged" in a few steps before it actually lands on production. The next section explains this promotion path.

Understanding the promotion path

Although eventually code will be pushed to production, it's good practice to run the deployment to a development , staging, or pre-production environment. This is called a promotion path, which involves **development**, **test**, **acceptance** and **production** (**DATP**). In some cases, an acceptance environment is part of the promotion path. Acceptance environments sit between the test and production systems.

This is shown in the following diagram:

Figure 2.5 – A promotion path for IT systems

Acceptance environments are strongly recommended for business-critical applications. In fact, enterprises use acceptance environments often as the **disaster recovery (DR)** system when production fails. Production is then switched to acceptance, with a minimal loss of data. Acceptance environments should be identical to production systems in terms of configuration.

A promotion path containing DTAP is recommended as a minimal setup. The executable package is the outcome of the integration pipeline and is now pushed to development, pre-production, and testing or staging. Here, the packages are tested to verify that they will run without issues in production. The application needs to be robust, stable, secured, and compliant at a minimum. Also, cloud-native features such as scalability—and maybe even self-healing—are tested, with load and performance tests. Ultimately, the final product will go live against the specifications that were originally collected from the business. We are then back at the beginning of the cycle, which started with demand management.

Why would we run such an intensive process to release code? Because there are several benefits a business could gain from this. The five main benefits are listed here:

- Code changes are more controlled and can be smaller since this is a continuous process where improvements can be made in small iterations of that code.

- Because of automation, code release can be done much faster.

- Because of automation, the costs for development will be significantly lower—this will be reflected in the overall costs to the enterprise.

- Automated tests will lead to more reliable tests and test results.

- Management and updates are easy.

In this section, we learned about CI: what the requirements are, how we initiate integration from automated pipelines, and why enterprises should implement CI in projects. There's one more thing we need to discuss as part of the full DevOps setup, and that is collaboration. DevOps is not only about technology and tools; it's a *mindset*. We will discuss collaboration in the last section of this chapter.

Designing and managing collaboration

To put it very simply, DevOps only succeeds if teams work together. Teams can collaborate if they use the same processes and, indeed, the same toolsets. In DevOps, **collaboration** ties processes and technology together to enable teams to join forces.

In *Chapter 1, Defining the Reference Architecture for Enterprise DevOps*, we saw that a lot of enterprises have outsourced major parts of their IT. This makes collaboration hard, every now and then. DevOps requires that teams carrying out operations and that are part of a certain sourcing partner or vendor work together with developers that come from a different company. It's up to the enterprise to set the scene, engagement rules, and co-working principles. The ownership of that can only be at an enterprise level.

In enterprises, it's very rare that only one team is completely responsible for an application. Often, there are more teams involved, and—even—more than one supplier. DevOps, however, assumes that there's one DevOps team responsible end to end. The main goal of this is to reduce overhead and manage waiting time in handovers between teams. This requires constant communication between team members since these members will have different skills and tasks.

Don't be fooled by the **T-shaped** dogma. Yes—a cloud engineer will probably know how to attach storage to a VM, and maybe they also know how to build network configurations in Azure or AWS, yet configuring databases or a firewall are specific tasks that typically require a database or a network engineer. A DevOps team will have to onboard these **subject-matter experts** (**SMEs**), but they work together in one team and not in separate silos.

The next thing is that an enterprise needs to enable collaboration. This is the task of a project leader or, in an agile way of working, a *Scrum* master. One of their key tasks is to "remove barriers" and help team members to achieve a common goal. Setting the objectives and priorities for the team are step one.

How do we define the objectives and the priorities? This is the task of an architect. Their responsibility is to lay out a clear roadmap, pointing out how to reach each objective, and in what order. That roadmap is derived from the end-state architecture, also referred to as "**soll**", or **target architecture**. This architecture defines how the environment should eventually look, containing the application architecture, the required infrastructure components, and the **application programming interfaces** (**APIs**).

An example of a roadmap is shown in the following diagram. The example has been taken from *ProductPlan* and shows perfectly how a simple roadmap helps create visibility and clarity in tasks that have been foreseen:

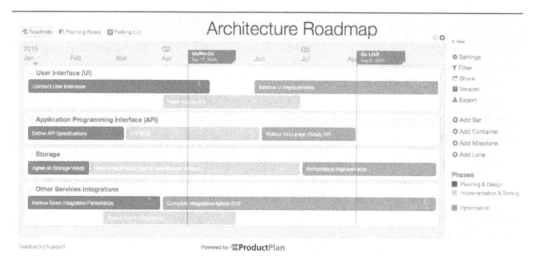

Figure 2.6 – Example of an architecture roadmap by ProductPlan

Finally, the roadmap shows the roles and responsibilities of team members: who needs to do what, and how much time will it consume? A good way to create clear overviews of roadmaps, timelines, and tasks is to use **Kanban boards**. Kanban, originally a Japanese invention that was first introduced at Toyota, shows when certain components are needed in the production process. Both Scrum and Kanban are tools to implement Agile as a development method for applications.

To a lot of enterprises, this still sounds very new, but the truth is that they probably have been working with a similar planning methodology for years, since that's what Kanban does in essence. It helps teams to plan complex projects in achievable chunks and, with that, improve the quality of the end product.

In the next chapter, *Architecting for DevOps Quality*, we will learn more about implementing quality measures.

Summary

This chapter started with an overview of demand management as input for an architecture. We learned that assessing business demand is the key driver for portfolios. In turn, a portfolio defines the artifacts and building blocks that we use to develop products, services, and, as such, applications. Since demand is changing fast, enterprises will need to speed up deployment processes. This can be achieved by automating as much as possible. Automation is done through pipelines, and in this chapter, we've learned what the different components are in architecting both application deployment and infrastructure pipelines.

In the last section, we discussed collaboration that is crucial in CI/CD, using DevOps. Teams will be formed by engineers with different skill sets, and they even may be hired from different companies, something we see frequently at large enterprises that have outsourced their IT. Therefore, enterprises will need to encourage strong collaboration, based on a clear roadmap with clear objectives and planning that states exactly who is responsible for which task. For this, Kanban boards are a good approach.

In the next chapter, we will dive deeper into the quality measures of DevOps projects and learn about **Definition of Ready** (**DoR**) and **Definition of Done** (**DoD**), as well as looking at how to execute tests for quality assurance.

Questions

1. Demand management is a process of collecting ideas and identifying opportunities for future products and services that are in a portfolio. Rate the following statement true or false: the business case is not relevant for demand management.

2. One of the first tests is a test without actually running the code. What do we call this test?

3. The promotion path contains various stages. What are these stages, and in what order are they set?

Further reading

- *Demand management as a critical success factor in portfolio management.* Paper presented at PMI® Global Congress 2016 by Romano, L., Grimaldi, R., & Colasuonno, F. S. (2016):

  ```
  https://www.pmi.org/learning/library/demand-management-
  success-factor-portfolio-10189
  ```

- *Kanban and Scrum – Making the Most of Both*, Henrik Kniberg and Mattias Skarin:

  ```
  https://www.infoq.com/minibooks/kanban-scrum-minibook/
  ```

3
Architecting for DevOps Quality

The overall aim of **DevOps** is to deliver high performance and quality to IT projects. In this chapter, you will learn how DevOps can add value to the quality of **IT delivery**. In this chapter, we will learn how to define test strategies, proving that quality has been delivered according to the **Definition of Done (DOD)**. But what happens if something breaks? The golden rule in DevOps is you build it, run it, break it, and then you fix it. But then, we must detect what the issue is by executing a **root cause analysis (RCA)**. In the final section, we will discuss **remediation** and, with that, **continuous improvement**.

After completing this chapter, you will be able to identify and implement quality measures in DevOps projects. You will have learned what tests can be included, how the are organized, and what the value of these tests is, thus continuously improving the product or service.

In this chapter, we're going to cover the following main topics:

- Defining test strategies
- Implementing quality measures
- Designing test automation and execution
- Understanding root cause analysis
- Designing for remediation

Defining test strategies

In the previous chapter, we concluded that testing is a crucial step in the CI/CD process to ensure the quality of the build. In this section, we will learn how to define test strategies in DevOps.

First, DevOps requires a different approach to testing: it's part of the continuous deployment and integration of *builds*. The reason we should adopt DevOps is because enterprises want to speed up releases so that they can act much quicker to changing demands. For testing, this means that there is a shift from testing the end product to continuous testing, with a focus on reducing build and test times.

In other words, testing is no longer just a matter of detecting the faults in the end product; it has become part of the full life cycle of the build, collecting feedback during this whole cycle.

Testers should be members of the DevOps team. Their responsibility is to constantly collect feedback, measure the cycle time, and find ways to reduce these times. Testers should be invited to monitor code during the whole build process so that they can look at errors and bugs as soon as they appear. This will give the team opportunities to fix issues before code is released to the next stage. The big benefit of this approach is that the end product will already have fewer issues and the total cycle time will be decreased during every iteration.

Overall, we can say that the role of quality assurance is changing in DevOps. In traditional IT projects, quality assurance was done as soon as the product – for instance, an application – was delivered at the final stage. Testers would execute functional testing and involve a group of users to perform **user acceptance tests** (**UAT**). This was done for a given period, usually a week or two. Then, the results were handed back to the developers that would go over the findings and fix the issues. Then, the whole thing was submitted to testers again so that they could retest and validate whether all the findings had been addressed. In DevOps, we don't work that way anymore. Firstly, this is because it takes too much time, and secondly, there's hardly any interaction between the testers and the developers.

What are the requirements for a test strategy in DevOps that ensures quality through the whole development cycle? They are as follows:

- **Create user story awareness**: First, there needs to be a clear **user story**. The user story will drive the test scenarios. This means that the team scopes the testing topics. **TMAP**, the most used testing framework, divides the topics into two categories: *organizing* and *performing*. Organizing topics cover the way tests are planned, prepared, and managed. Performing topics concern the tests themselves.

- **Create the strategy**: In DevOps, the **strategy** should be that tests are executed throughout the whole build, which means that testers need to run automated test scripts on iterations of the build. In other words, code is constantly tested to verify it performs well without issues. This requires strong cooperation between the developers and the testers: during the build, developers will have to supply code to testers, as well as to interim builds, until the code is stable.

  Obviously, there has to be a balance between the number of tests, the testing time, and the goals of the tests. Enterprises run tests to avoid risks such as materializing and causing damage. One of the first topics that needs to be addressed in defining a test strategy is creating a clear view of risks.

  Be aware that this is not just something technical. It also involves soft skills in collaboration and communication. Where testers were used to get the whole package delivered at the end of the development cycle and then run their tests, they are now part of the DevOps team, continuously communicating about timing, the way of testing, and what will be tested. This is crucial for the success of DevOps.

- **Define tools and environments**: Honestly, it's not about the tools – it's about the code and the level of **standardization**. Testers need to make sure they capture all the code: we refer to this as code coverage, which should be 100% or very near that at the least. Test cases must be automated. **Automation** requires standardization: code needs to be automatically deployed to standardized test environments where the pre-testing, cleanup, and post-testing tasks are automated. This will increase efficiency and reliability, preventing human errors in repetitive tasks, assuming that a number of tests will be executed more than once.

  A first test, very early in the development process, includes *static analysis* to check whether the code is complete. With static analysis, the code is not executed: tools validate that the code is complete and consistent. An example is when testers can use tools and scripts to validate that security policies have been applied to the code and that the code is compliant with industry standards. Reviewing the static analysis process should provide a detailed overview of the quality of the code and the surrounding documentation.

- **Execute**: Test scenarios must be very well structured. This needs to be defined in the test design, which we will discuss in the *Designing test automation and execution* section. There, we will look at various executing techniques:

  - Process focused

  - Condition focused

- Data focused

- Experience focused

An important goal of DevOps is to speed up delivery by reducing waiting times between process steps. This also includes testing. We already noted that testers do not have to wait until the code is finally delivered, but that they can run automated tests in subsequent iterations of the build cycle. To reduce the test time – and with that, the whole build – *parallel execution* of tests is advised.

There's more that can be done to improve quality and tests and that's not the sole responsibility of quality engineers and testers. DevOps really is a team effort, encouraging all members to contribute to various steps and stages. Developers are therefore also invited to contribute and add test cases. It's good practice to collect cases, scripts, runbooks, reports, and other documentation in a quality repository. From this repository, quality engineers and testers can pick up cases and further automate these in the development and deployment process.

- **Creating and interpreting reports**: Test results must be evaluated. At the beginning of this section, we stated that tests are executed to identify, analyze, and mitigate risks. Test outcomes must match these risks, meaning that they should result in expected outcomes. If the outcomes are completely different, then the risks need to be investigated further. Only then will the tests really contribute to the quality.

  So, testing is much more than just detecting bugs. Tests must be focused on the overall desired outcome of the build. Consequently, testers need to report on much more than purely the issues. Reporting is now really focusing on improving the quality of both the build and the build process. An example is when the test results may show where automation can be improved or code can be enhanced. The overall goal is to reduce risks.

  Blocking issues – issues that inflict a big risk – must be reported instantly and looped back to the start of the development chain.

- **Setting exit criteria**: Results matter in tests and they lead to decisions about if and how to proceed. That's what **exit criteria** are for. If all the necessary tests have been conducted, the results should give you enough information to do a *go/no go* and push the software to the next stage, typically production.

In this section, you learned how to compose a quality and test strategy. Before we learn how to implement quality measures, we will briefly look at different types of tests.

Understanding types of tests

In this section, we will introduce the different, most commonly known types of tests. There are three tiers of testing:

- **Level 1**: **Small tests** that focus on separate components. Unit tests are an example where small pieces of the code are tested.

- **Level 2**: **Integration tests** that involve more than one component. Integration itself is tested, but also how components interact and if the integrated packages perform well. Integration and performance tests are executed at this level.

- **Level 3**: **Usability tests** that focus on how easy the end product can be used – for example, using the graphical interface – and if it's easy to manage. The **User Acceptance Test (UAT)** is typically the final test at this level. To be clear, the UAT also involves performance testing from the end user's perspective.

These three levels are shown in the following diagram:

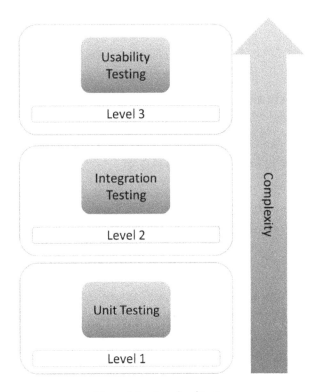

Figure 3.1 – Levels of testing

The complexity increases as testing moves from single components to a usability test, which involves the full stack of the solution.

In the test strategy, the teams will define, based on the business requirements, what tests must be conducted and what the expected outcome must be to deploy a build successfully.

A couple of specific tests have not been mentioned yet:

- **Regression test**: These tests were very common in the traditional approach of software development. Functional and technical tests were re-executed to make sure that software – code – that was changed during the development cycle still performed without issues after the change. As we have seen, DevOps has changed the way we approach testing. It's not a one-time exercise anymore, where we run a regression test, find bugs, fix these, and rerun the tests. In DevOps, code is continuously tested and improved throughout the build. Regression tests have become *less important*. In some cases, it still might be valuable to execute regression tests.

- **Security test**: The same applies to security tests that were often executed once the build was delivered. In DevOps, we check for vulnerabilities and compliancy issues during the first static analysis stage. In *Chapter 14, Integrating DevSecOps with DevOps*, we will go into this in more detail.

Testing is about validating quality. In the next section, we will learn about quality measures and how to implement them in our DevOps projects.

Implementing quality measures

By now, it should be clear that everything in DevOps is about being **continuous**, which, in other words, means **continuous deployment**, **continuous integration**, **continuous testing**, and **continuous quality** engineering. DevOps projects constantly focus on **quality** at every stage of development and operations. It's different from traditional approaches where teams have a separate phase to fix issues. In DevOps, teams constantly measure the products and fix issues as soon as they occur. One of the six DevOps principles is **continuous improvement**, which refers to the feedback loop wherein products are improved in every iteration, but also to the DevOps process itself.

A common practice in IT projects was to have a fixing phase, something that Gerald Weinberg describes in his book *Perfect Software and other illusions about testing*. The fixing phase was put at the end of the development phase before software was handed over to operations. In DevOps, we don't have a fixing phase because quality is measured and tested as teams go along, during the whole development and operations cycle. This can be seen in the following diagram:

Figure 3.2 – Continuous testing (based on TMAP)

How do you measure quality? First, what is quality? We started this book by saying that it all starts with the business. That's why **enterprise architects** have such an important role to play in DevOps. Enterprise architects have a major task in translating business needs into solutions. From business demand, the product portfolio should be clear about what standards these products should be delivered to. That's quality. It's a product or service that meets the needs and expectations of the user. Quality is a product or service that satisfies the users. Hence, testing is about validating if the user can be satisfied with the delivered product or service.

What do enterprises measure in terms of DevOps quality? Two main things: the build itself and its *functional validation*. For the build itself, topics such as the number of successful builds, the total number of defects, and code coverage are important. For functional validation, the main topics are whether all the requirements have been tested and if all the identified risks have been covered.

In the next few sections, we will learn what instruments we can use to define quality and how to measure it, starting with acceptance criteria.

Defining acceptance criteria

Before we can start testing, we need to know what we will be testing. That's why we need to define **acceptance criteria**. To put it very simply: when is good, *good enough?* Again, it starts with the user story, which defines what a product or service should include. The user story sets the scope and the specific functionality a product or service must have.

How do you set acceptance criteria? The question is, *what is it that we're building?* "It" must be specified and with that, "it" must become tangible. The DevOps team looks at specifications from four angles:

- **Business**: What does the business require?
- **Development**: How can we build a solution that meets the business requirements?

- **Quality**: How can we test the solution and validate that it meets the requirements?

- **Operations**: How can we manage the solution so that it keeps meeting these requirements?

In DevOps, we're not building the whole package at once, as in a waterfall type of project. The team starts with a minimal viable product and then iterates the product, continuously improving it. Acceptance criteria are set per requirement and thus, following the logic of DevOps and quality measures, each requirement is tested. This is what **test-driven development** (**TTD**) does. In TDD, the team writes the test case first and then the code. The code is written to the specifications of the test case, proving that the requirements have been met. The TDD flow is shown in the following diagram:

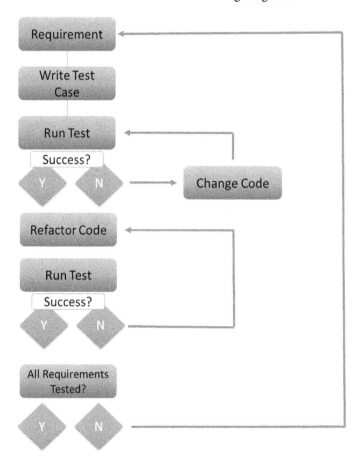

Figure 3.3 – Test-driven development

TDD is nothing new and has existed since the mid-fifties, but the most commonly used version is described by Kent Beck (refer to the *Further reading* section). The team picks a requirement, writes the test case, develops the code, and runs the test. When the tests are successful, the code is refactored or rewritten, meaning that the code is cleaned and remediated to the standards the architect has set for the code. After this step, the code is tested once more with the test case. This cycle is repeated for every requirement.

The next steps are assessing the Definition of Ready and agreeing on the Definition of Done. We will study them in the next section.

Defining the Definition of Ready and Definition of Done

In the previous section, we learned how to set acceptance criteria and how TDD can help in making sure that we meet a specific requirement. However, a product or service likely has many requirements before it can really be launched into production.

In DevOps, we use two important processes to validate if a fully developed product or service is ready for production. These processes are the **Definition of Ready (DoR)** and the **Definition of Done (DoD)**.

To avoid mistakes, bear in mind that the acceptance criteria are not the same in both DoR and DoD:

- **Definition of Ready**: To understand DoR, it's important to know how *Agile Scrum* works. DevOps teams typically work in sprints, a short period of time in which a piece of the product is developed. The work that needs to be done is defined in **product backlog items (PBIs)**. The whole product or service is defined in a user story and then broken down into PBIs – tasks that a team can work on during a specific sprint and that can be completed during that sprint.

  Agile Scrum actually doesn't mention DoR. Yet, it has become a common practice to have a set of agreements to define when PBIs are ready to be picked up by the team. The problem with user stories is that, in some cases, they don't contain enough concrete information to start the build. The DoR contains entry criteria so that the team knows what to build. The process of defining the DoR is referred to as refinement.

- **Definition of Done**: In contrast to DoR, the DoD is part of Agile Scrum and describes exactly what a **Product Backlog Items (PBI)** looks like when it's finished. So, the DoD is a very strong instrument for validating the quality of the builds. Developers commit to the DoD. They commit to the fact that they must be absolutely clear about what they need to build. In IT projects, the DoD must contain the following topics as a minimum:

  - All the code has been written.

  - All the code has been reviewed and validated.

  - The relevant documentation has been written and made available.

  - All the tests have been executed and signed off.

  - All functionality has been proven to be delivered.

  Where the DoR takes care of the entry criteria, the DoD contains the exit criteria – the statement of completion. All team members must agree upon the DoD: the business, developers, testers, and operations. Don't forget the last group, where operations need to sign off the DoD and run the software. For them, it's crucial to validate that the code and relevant documentation are completed. Furthermore, in true DevOps, we will not have separate developers, testers, or members doing operations. Instead, we will have members with skills in development or testing.

So far, we have discussed the test strategy, the quality measures, and the acceptance criteria. In the next section, we will learn how to design test automation and execution.

Designing test automation and execution

In this section, we will learn how to design and implement tests. We will study the most common different test varieties and learn where we can use them. When we start discussing testing in IT, we need to discuss and agree upon a test management approach. In this book, we will use TMAP, introduced by ICT service provider Sogeti in 1995 and widely accepted as the standard in software testing.

The traditional phases of TMAP are as follows:

- Planning
- Preparation
- Specification
- Execution
- Evaluation

In DevOps, this is not a one-off exercise; we will be working with continuous testing. One major difference with the traditional way of working is that there's no separate test unit with a manager and testers. These professionals are now part of the DevOps team and they do their work alongside the developers. Next, in DevOps, we are working according to the *everything as code* principle, thus allowing teams to automate as much as possible.

Before we learn more about continuous testing, we need to understand the various types of testing. The most important ones are as follows:

- **Process focused**: The test focuses on the paths and flows in software. In TMAP, this is called **path coverage**. The test covers the path that a transaction follows to reach a certain end stage. This can become very complicated when lots of paths need to be followed. In that case, the test covers all possible paths and various decision points.

- **Condition focused**: A condition can be a decision point. The test covers the decision points and the conditions that it points to be either *true* or *false*. The question is how that will influence the outcome of the test. Be aware that this is a very simple explanation. In theory, software will have many decision points with specific conditions, thus influencing the outcome of the test. In a **Condition Decision Coverage** (**CDC**) test, all decision points and decision outcomes will be tested against the condition *true* or *false*.

- **Data focused**: This test uses data input to verify the test results. This type of test is commonly done through **boundary value analysis** (**BVA**). In the test, we enter the minimum and the maximum values, known as the *boundaries*. These can be numeric, but also *statements*. If we then, as a test case, enter a value that's outside these boundaries, the result should be "invalid." Any input within the range that is specified should lead to the result "valid."

- **Experience focused**: These are more often referred to as **user experience (UX)** tests. All of the previous tests are binary. The software follows the expected paths, the conditions of the decision points lead to expected results, and the entered data gives the expected results. Experience is something completely different since it's very subjective by nature. Yet, testers would like to know how the software "feels." Is it responsive enough, does it perform well, and is it easy to use? The basic question is: *is there a way to test experience in an objective manner?* There are some methodologies that aim for this, with one of them being the "experience honeycomb," which was developed in 2004 by Peter Morville. This can be seen in the following diagram:

Figure 3.4 – The experience honeycomb

Still, experience remains a bit intangible. It's very useful for finding out whether software is meeting the user's expectations, but to find faults and issues in the software, testers will need to perform more exact tests. Note that experience-oriented tests are very hard to automate.

Again, testing in DevOps is not a one-off. In the next section, we will discuss continuous testing.

Understanding the principles of continuous testing

In *Chapter 2, Managing DevOps from Architecture*, we learned about the CI/CD pipeline. We saw that testing was an integrated part of the pipeline. This assumes that all software is developed using CI/CD, which, in enterprises, is often not the case. There will still be, as an example, legacy systems that are not integrated in CI/CD. The same applies to **Software as a Service** (**SaaS**) applications: these are purchased as a service, and for that reason, not "developed" within the enterprise. However, they need to be tested.

The first step in continuous testing is automating the **test cases**. This is easier said than done, but it's feasible if tests have been set up from known patterns – for instance, simulating how a user would use the application. If we have an application that processes purchases, we could think of three use patterns:

- Place an order
- Cancel an order
- Check the order's status

Steps can be automated and with that, we can create a test case that can be executed.

Once we have the test cases and the code that needs to be tested, we need an environment where we can execute the tests. This can also be automated. By using the public cloud, it's easy to create a (temporary) test environment on demand and decommission it automatically when the tests have run. It could even be part of the full test scenario, where you can spin up an environment in Azure, AWS, or any other cloud, deploy the code (or rather, a copy of the source code), run the tests, and decommission the environment after completion. This does require automated infrastructure to set up and **Infrastructure as Code** (**IaC**) to be implemented, something that we will discuss later in this section.

In summary, continuous testing requires the following:

- *Integrated quality engineering and testing*: Testing is an integrated part of the DevOps team – meaning that every member of the team is involved in testing. However, in large enterprises running multiple DevOps projects, it is recommended to have a quality team that helps implement quality measures and tests the strategies in these projects.

- *Automated test cases*: It's recommended to start small. Pick one test case and automate that. Set the baseline for this test case: what data is absolutely required to run a successful test? What metrics will be used and for how long will a test need to run? Don't make tests too big; use small test sets and run them for a short period of time. Evaluate and, if needed, adjust the test sets and their duration.

- *Test tools*: These tools need to integrate with the CI/CD pipelines. This book is not about test tooling, but some popular tools are **Selenium** and **Micro Focus Unified Functional Testing** (**UFT**). How do you pick the right tools? That really depends on your approach. Some enterprises prefer a single stack solution, meaning that one tool covers the whole testing strategy. Others have a toolset, using separate tools for test modeling and test execution. Again, integration with the CI/CD pipeline is crucial.

- *Automated test environments*: Automate how the test data is provisioned, how the test case is executed, and how the test environments are provisioned using cloud services. Automating environments requires that we define everything as code:

  a) **Infrastructure as Code** (**IaC**): Virtual machines, network components, containers, firewalls – everything is defined as code. In Azure, we use **Azure Resource Manager** (**ARM**) templates, and in AWS, the preferred way of working is to use **CloudFormation**.

  b) **Configuration as Code** (**CaC**): As soon as infrastructure components are deployed, they need to be configured so that they match the standards and policies of the enterprise. Think of DNS settings, certificates, and security policies such as hardening and patching. Once the configuration has been applied, we reach the **Desired State Configuration** (**DSC**).

  > **Important Note**
  > DSC is a term that is typically associated with Microsoft Azure. In this book, we will use DSC as a generic term to explain that the cloud infrastructure needs to meet specific requirements, as defined by the architecture, in order to be compliant.

  c) **Pipeline as Code** (**PaC**): Every step in the CI/CD process is defined in code, from pulling the source code to its actual deployment, including the test procedures.

  d) **Test as Code** (**TaC**): Test as code refers to the test automation process itself, from collecting, assessing, and deploying test data to actually executing (running) the various tests and collecting the results. We can also validate the results automatically using *artificial intelligence* and *machine learning*.

In this section, we learned all about testing as a methodology to validate quality in DevOps projects. We saw that automation can bring a lot of value to our testing strategy. One important remark must be made as a conclusion: *it's not about automation itself*. The goal should be to optimize the builds and improve their quality. It's the quality value that matters, not the test itself. Tests will help teams improve the quality of their work by identifying risks and helping them understand how to mitigate issues. That's what we will talk about in the next section. We've found a problem – *now what?*

Understanding root cause analysis

In the previous sections, we discussed quality measures and testing to validate these criteria in a highly structured and automated way. Still, things can go wrong. The golden rule in DevOps is *you build it, you run it*, often followed by the statement *you break it, you fix it*. Or even it could be *you destroy it, you rebuild it better*. If something breaks, the team will need to find out what exactly happened. In this section, we will talk about **root cause analysis** (**RCA**) as one of the most important instruments for finding the cause of a problem.

RCA is the methodology for finding the exact cause of an issue. With that, RCA provides insights on how the team can improve products or services. These can be quick fixes or long-term enhancements. RCA is more than just a way to find problems; it's the start of improvement. Important questions that need to be addressed in RCA are as follows:

- What is the problem?
- Where was it found?
- Why did the problem occur?
- What caused the problem?
- What improvements can we make to avoid the problem in the future?

There are several ways to conduct an RCA. Popular methods are the *5 whys* and the *fishbone* diagram (also known as the *Ishikawa* diagram). The 5 whys method is an easy way to literally drill down to the root cause of a problem, simply by asking "why" five times. It's a bit like a little child that constantly repeats the same question up until the point where it's satisfied with the answer.

The fishbone, invented by Professor Ishikawa, is more suitable for drilling down to more complex issues. Again, it starts with the problem and then the team identifies what could contribute to that problem: infrastructure, code, programmers, and so on. These are the "fish bones." Each of the bones is then analyzed. The basic diagram for this is as follows:

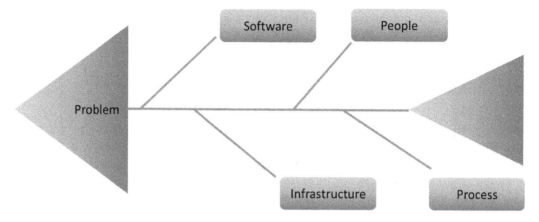

Figure 3.5 – Fishbone diagram

Regardless of the methodology, the basic steps for RCA are always the same, as shown in the following diagram:

Figure 3.6 – Steps of RCA

The RCA starts by gathering data to find out what exactly happened. The next step is the problem statement: when did it happen, where, and what is the impact of the problem? The last question in particular – *what is the impact?* – is important. It drives prioritization in the project and the business case. If the impact is low but the mitigation solution will require a huge investment in time and thus costs, the team might decide to give it a low priority and put it on the backlog of the project. If the problem has a high impact, it might become an impediment. It needs to be solved before the team picks up any new tasks. It's one of the main principles in *Site Reliability Engineer*, which we will discuss in detail in *Chapter 5, Architecting Next-Level DevOps with SRE*.

After analyzing the cause and the impact, the team can work on solutions to mitigate the problem. The last step is the final report. It's a common practice to test the solution first and validate if the solution is really solving the problem. RCA is a quality measure and quality measures need to be tested, as we learned in the previous sections.

With that, we have discussed testing and how to handle RCA to improve the product by finding solutions for problems. But as with everything in DevOps, we are aiming for continuous improvement. That also includes roadmaps for improving the building blocks themselves, known as the infrastructure, the coding framework, and the DevOps environment. That's the topic of the next section.

Designing for remediation

So far, we've talked about coding the software, implementing the required infrastructure, automating it all through CI/CD pipelines, testing the environments, detecting issues, and, if needed, fixing the problems. But there's something that we haven't been discussing yet and that's the speed of software development and DevOps itself.

DevOps is about learning. As the team and projects grow, they learn how to improve. They learn from the product itself and how it's used, and they learn from looking at other projects, technologies, and methodologies. These lessons are adopted and injected into their own project. The team doesn't need to start over, though – they can adopt and adapt as they proceed. We call this **remediation**, which is the process of improving an existing situation.

Remediation can take place on three levels, as follows:

- *Infrastructure*: Assuming that we build everyone according to the "everything as code" principles in public clouds such as Azure or AWS, teams will have to take into account that these platforms evolve rapidly. It's the responsibility of the architect to "track" the roadmap of the cloud services, and then decide whether to include new features in the roadmap of the project and improve the infrastructure.

- *Software/application code*: Software developers work in code frameworks or versions, such as *.NET*. The framework contains the **Framework Class Library** (**FCL**), which holds the languages that code can be written in to ensure interoperability between different platforms. By using compilers, code written in C#, VB.net, and J# (Java) is translated into **Common Language Infrastructure** (**CLI**) so that it runs on Windows platforms without us having to write machine code directly. CIL produces executables that can run on Windows and various Linux distributions such as **Red Hat Enterprise Linux** (**RHEL**), Ubuntu, Debian, Fedora, CentOS, Oracle, and SUSE. .NET is just one example. Other frameworks include ASP.NET, Java, Python, PHP, and JavaScript. They all run specific versions and developers must make sure that their code is running a supported version. Again, it's recommended to have the framework versions set out in a technology roadmap to keep track of the life cycles.

- *DevOps*: Finally, DevOps itself has various implementations, typically in combination with a specific way of agile working. In other words, it's not only the tool or toolsets that change, although it's important to keep track of the DevOps tool roadmaps. It's crucial for source control. For example, Azure DevOps – widely used to run DevOps projects in Azure – currently runs Azure DevOps Server 2020 as a version control system, allowing developers to work together on code and track changes.

The key takeaway from this section is basically to never stop learning and never stop improving. IT is changing rapidly and so is DevOps. DevOps teams have a great responsibility in staying ahead so that the business really can benefit from new developments. It's the architect that has the responsibility of guiding the teams in this and making the right decisions. With that, the architect should focus on *quality*.

Summary

This chapter was all about quality. We learned how to identify quality measures and that this is more than just about fixing bugs. Quality is about meeting expectations, but DevOps teams need to be able to measure these expectations. That's why businesses, developers, and operators need to be clear on the acceptance criteria. In this chapter, we discussed the DoR as an entry point to working on a project and DoD for measuring whether a product is really complete.

Measuring means that teams have to test. In a traditional way of working, testing is done as soon as the whole product is delivered. In DevOps, we work with continuous testing. In other words, all the team members are involved in testing and validating the quality of the product. In this chapter, we discussed different ways and types of testing that are common in DevOps. Lastly, we talked about continuous improvement using remediation. Cloud platforms, software development technology, and DevOps tools are constantly evolving, and DevOps teams need to adapt and adopt these changes in their projects to allow businesses to benefit.

The role of the architect is crucial in that they need to guide in these developments and enable the team to make the right decisions.

In the next chapter, we will discuss scaling DevOps. We start small, but in enterprises, we need to scale out if we want an entire business to start working agile and work in DevOps teams. On this note, what do we do with existing programs and projects? Let's find out!

Questions

1. In this chapter, we discussed different types of tests. One of them is unit tests. Give a short description of a unit test.

2. In a data-oriented test, we enter the minimum and the maximum values. If we enter a value within these boundaries, the test result should be valid. What is this test method called?

3. To decide if a product is complete, DevOps uses a certain technique. What is this technique called?

4. *True or False*: A fishbone diagram is a good practice for analyzing the root cause of a problem.

Further reading

- *Quality for DevOps Teams*, by Rik Marselis, Berend van Veenendaal, Dennis Geurts and Wouter Ruigrok, Sogeti, 2020

- *Test-Driven Development: By Example*, by Kent Beck, 2002

4
Scaling DevOps

DevOps started—and is in some companies still done—with a lot of manual tasks, scripts, and ad hoc tests. A lot of enterprises focus on the applications and tend to forget about the platform itself—the infrastructure—but this is also crucial to the scaling. This chapter focuses on the scaling of DevOps, both from a technical and an organizational perspective.

After completing this chapter, you will have learned how to handle scaling. First, we will learn about modern DevOps, which adopts cloud and cloud-native technology as target platforms to run applications. Before we can do that, we probably need to transform the applications; otherwise, we will develop new applications. In DevOps, we need a development method that fits to the way of working; therefore, we will discuss **rapid-application development** (**RAD**). Next, we will look at adopting DevOps throughout a whole enterprise, starting small and then expanding. Finally, we will have a look at mission-critical environments and how we can manage them in a DevOps mode.

In this chapter, we're going to cover the following main topics:

- Understanding modern DevOps
- Working with RAD
- Scaling infrastructure with DevOps
- Scaling DevOps in enterprise environments
- Managing mission-critical environments with DevOps

Understanding modern DevOps

The concept of DevOps is not new. Basically, the idea was that teams could improve their work if developers and operators were really working together as one team. The reason for that was easy to find, as we've already seen in *Chapter 1, Defining the Reference Architecture for Enterprise DevOps*. In this section, we will learn how DevOps has evolved over the years and what the impact of modern DevOps has been on enterprise **information technology** (IT). We will also study how DevOps helps in transforming legacy applications by app modernization.

A lot of enterprises decided in the 1990s that IT was not a core business and could be outsourced to suppliers. Typically, all management—operations—of commodity IT was outsourced. It created not only silos within enterprises, but also outside of them. Over time, IT got more complex, demands increased, and enterprises found themselves in a position of having to find ways to get back into the driving seat of IT, which had become core for the business.

In 2009, the first **DevOps Days** conference was held in Belgium. The base idea: break down the silos, put developers and operations back in one team, and increase the quality and velocity of software development. These are still the goals of DevOps. That hasn't changed over the years.

But it doesn't mean that DevOps hasn't changed at all. The main differentiators are cloud technology and automation. These are the two most important pillars of modern DevOps and are outlined as follows:

- **Cloud**: One of the main issues in early DevOps was the availability of development and test systems. In modern cloud platform deployments, even temporary systems have become much easier.

- **Cloud native**: Silos between developers and operations have been broken down, but the same applies for technical silos in different platforms. Interoperability between systems has become the standard in cloud native, with the entrance and emergence of **Platform as a Service** (**PaaS**), **Software as a Service** (**SaaS**), container technology, and serverless functions. The major two developments in IT for the near future are likely software to services and **virtual machines** (**VMs**) to containers, increasing portability across cloud platforms and even between on-premises systems and cloud systems.

- **Automation**: Automate as much as possible. This means that in modern DevOps, we should perceive everything as code—not only the application code, but infrastructure, configuration, and integration as well. In *Chapter 2, Managing DevOps from Architecture*, we discussed pipelines for the deployment of applications and infrastructure, but in modern DevOps everything is built through pipelines.

- **All code is stored in a repository**: Applications, infrastructure, tools, governance, and security are translated into code, and with everything as code, we can have everything in pipelines. So, besides a deployment pipeline for application code and a pipeline for infrastructure, we will also have pipelines for tool configurations, integrations, reports, and security.

- **Integrated security**: For security, we will have DevSecOps, which starts with Security as Code. The security posture of enterprises is translated into code and managed from a single repository. The development of the security posture is handled in the same way as the development of applications and infrastructure, by continuous improvement and not solely by reacting to attacks, threats, or breaches. Code that has been developed will immediately be merged with the security code, integrating the security posture. Security is developed at the same speed as applications and infrastructure. *Section 3* of this book, *Bridging Security with DevSecOps*, is entirely about implementing DevSecOps, starting with *Chapter 12*, *Architecting for DevSecOps*.

- **Enhanced technology**: Modern DevOps is sometimes referred to as *accelerated* or intelligent DevOps. With enhanced technologies such as **artificial intelligence (AI)**, **machine learning (ML)**, and **robotic process automation (RPA)**, automation can really be leveraged. Examples are self-healing systems or code and pipelines that *learn* autonomously from previous deployments. With RPA, processes can be highly automated and, if combined with AI/ML, think about logical next steps in deployments—for instance, by learning from test results or system behavior. **Artificial Intelligence IT Operations (AIOps)** is a good example of this development. *Chapter 8*, *Architecting AIOps,* and *Chapter 9*, *Integrating AIOps in DevOps*, will provide deep dives into AIOps.

To summarize, modern DevOps is more about this:

- Involving all stakeholders—it's not only about developers and operations, but also about business managers, security specialists, quality and assurance managers, and procurement (think of licenses).

- Cloud and cloud-native adoption, including containers, functions, and automated services.

- Thinking in code, and therefore in pipelines. Keep in mind that with everything as code and deep automation, we also need to think about the principle of trust. We need to make sure that the code and assets we have in our pipelines are trusted. Next, who is mandated to state that assets are trusted and may be applied to the pipeline? Is that the security engineer, or can it be delegated to the developer? Or, can delegation even be automated if we have systems that adhere to the **principle of least privilege (POLP)**? Segregation of duties becomes very important—controls are required to protect code from unauthorized changes.

In this section, we discussed changes to DevOps over the years. The reason for enterprises adapting DevOps is to modernize IT as a whole. Enterprises have a long history (also in IT), and therefore typically have complex, large IT ecosystems with legacy applications. App modernization has become an important topic in modern DevOps, and the next section will talk about that.

Introducing and understanding app modernization

DevOps, the cloud, automation, and code are all key principles of digital transformation. But many enterprises will have core applications that they have been running for a long time: legacy systems not fit to adopt DevOps or even ready to migrate to the cloud. In this section, we will discuss the process of application modernization: the process of transforming these applications to systems that we can run in the cloud, keeping up to date with modern technologies, and supporting the business by being able to adapt to new demands faster.

> **Note**
>
> App modernization is a huge market. A number of companies such as IBM and Fujitsu have massive programs to transform mainframe applications to cloud providers such as **Amazon Web Services (AWS)**, even running **common business-oriented language (COBOL)**. The reason for enterprises *shifting* their old mainframe applications to the cloud is easy to understand. The original code is left intact as much as possible, but it's moved from expensive on-premises equipment to pay-as-you-go cloud environments, where risks are fairly low. The downside is that companies will still need to have resources with COBOL programming skills to maintain the app itself. In other words, the application itself is still legacy and will probably not be able to adopt and benefit from new cloud-native technology. The next step would be to also modernize the application by, for instance, rewriting the code or rebuilding the functionality in a completely new application.

Enterprises can follow a number of strategies in app modernization, such as the following ones:

- **Rehosting**: This is the lift and shift of existing systems *as they are*. The application is picked up and moved to another target platform. The application is not changed in any way whatsoever. However, moving an application from—for instance—an on-premises environment to AWS or Azure will imply some modifications, especially in connectivity. These modifications will be very minimal and will not impact the application itself. A common way of rehosting is exporting the VM, including the operating system and application code as a whole, to the new cloud environment. It's a fast way to migrate applications to the cloud, but by doing so, enterprises will not experience real benefits from cloud services. They simply run their machines in another data center—in this case, the cloud data centers of AWS, Azure, or any other public cloud provider.

- **Replatforming**: With replatforming, applications are optimized to run in the cloud. A very common example is replatforming databases to PaaS. Database instances are moved to AWS **Relational Database Service** (**RDS**) or Azure **Structured Query Language** (**SQL**), both native database services managed by the platforms. Developers will still be able to program databases the way they're used to, but they don't need to worry anymore about the database platform itself. That is taken care of by the provider.

- **Refactoring**: Sometimes referred to as rearchitecturing. In this case, the application code is modified to run in an optimized way in the cloud. Cloud-native services are applied. Application code might be rewritten to run in containers or with serverless functions. The functionality of the application remains as it is: only the underlying technology is changed. Take the COBOL example at the beginning of this section: COBOL could be rewritten to C# or Java. However, architects and engineers would first need to decouple the business logic from the code itself. If the business requires the business logic to be modified as well, then the strategy changes to rebuilding.

- **Rebuilding**: In the case of a rebuild, architects start with revalidating the functionality of the application. What is the functionality that the business requires, which data will be used, and how does it translate into the usage of an application? Next, the application is rebuilt on the validated business and technical requirements. The functionality returns, the technology is completely revisited, and the application is rebuilt.

Now, rehosting will not bring real benefits in terms of costs. Cost savings might be achieved when applications are replatformed or refactored, since the use of cloud resources will be optimized.

Rebuilding is a different story. This might lead to major projects that will inflict project costs. As soon as the application is ready to be pushed to production in the cloud, substantial savings might be achieved. However, enterprises need to take the total cost of ownership into account, thus including the project costs and possible risks in rebuilding an application. Enterprise architects play an important role in advising and supporting decisions. The following screenshot shows a very simple overview of cloud cost components:

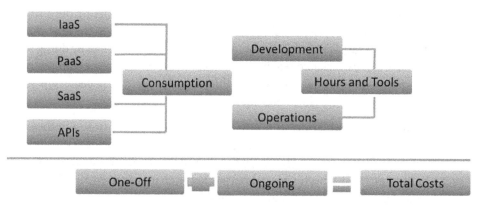

Figure 4.1 – Simple overview of cloud and DevOps costs

What are the steps an enterprise—and a responsible architect—should take to start with app modernization? Let's have a look at these here:

- **Importing**: The architect gathers all relevant data about the applications. They can do so through analyzing information from the **configuration management database (CMDB)** by using tools to scan the applications, and through workshops with stakeholders such as business and application owners.

- **Assessing and architecting**: The next step is assessing all the data. What does the architecture look like and how could it map to a modern—cloud—architecture? At this stage, the target architecture is defined as well as the *future mode of operations*, meaning the way the application is executed and managed. The DevOps mode of working and the use of **continuous integration/continuous development (CI/CD)** pipelines are included in the future architecture. This defines the method of transformation. In short: at this stage, the architect defines the *what* (what the application and the architecture will look like) and the *how* (how we transform the current application to a modern app). The *what* and *how* together form a solution.

- **Deciding**: Stakeholders are informed on all relevant aspects: functionality of the application, technical realization, risks, and costs. The business case is validated and, based on this, a go/no-go decision is taken to proceed or not.

- **Executing**: The project starts. Deliverables and technical roadmaps have been defined in features, product backlog items, and tasks. Components are refined and pulled into build sprints. Tests are executed, acceptance criteria are validated, and **definitions of done** (**DoDs**) are signed off as the project evolves.

The following diagram shows the high-level process of app modernization:

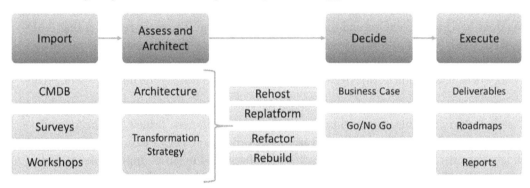

Figure 4.2 – High-level process of app modernization

In summary, app modernization is about the following:

- A compelling business case. Is it worthwhile to modernize an application? How *core* is the application to the business?

- A compelling strategy. Replatform, refactor, or rebuild? Or, is it wise to do a simple lift and shift to the cloud first and start the transformation on the new platform?

- A compelling plan. Have risks been identified? Does the team have the right skills and tools to mitigate these risks? Knowing the risks, is the plan feasible in a number of sprints?

We've discussed legacy applications and what enterprises can do to modernize these applications, but enterprises will also develop new code and launch new or improved services. Since we are working in a DevOps mode, we have to look at a development methodology that keeps track with that. RAD is a solution. We will learn about RAD in the next section.

Working with RAD

So far, we discussed how DevOps breaks down silos between developers and operations and how it helps in speeding up the development of products, services, and systems. Implementing DevOps will increase the velocity of development, but DevOps in itself is merely a way of structuring planning of development. It helps in planning in iterations: starting with a **minimal viable product** (**MVP**) and then iterating improvements in next versions. DevOps is not about the development of code itself. We need a development methodology for writing the code, but that methodology should *fit* with DevOps. In this section, we will look at RAD.

Why does RAD fit to DevOps and the agile way of working? The main reason is that RAD is agile in itself. RAD starts with prototyping (the MVP) and then focuses on iterations. The emphasis is on the fulfillment of requirements, rather than on planning. It allows developers to realize quick improvements and adjustments during the development cycle.

Key principles in RAD are furthermore reuse of code and intensive collaboration between the stakeholders: business representatives, architects, developers, testers, engineers, and the end customer that will use the software. Code is constantly reviewed, tested, and validated against the requirements, which are implemented in small improvements. This way, risks that the end product is not meeting the specifications are less likely to occur. The team is in full control of every single small step.

To develop according to RAD, the team needs to follow the following five basic steps. These steps completely align with the principles of DevOps:

1. **Define requirements**: Gather the business requirements and set the scope, budget, timelines, and the acceptance criteria. Have all stakeholders sign off to ensure that everyone is in agreement of the deliverables and the final product.

2. **Build**: The development starts with the MVP. Next, the MVP is improved in iterations up until the final product is delivered. Keep in mind that in DevOps, developers and operations need to be aligned on the product, so they will have to work closely together. Can operations manage the application or can it be improved? In *Chapter 5, Architecting Next-Level DevOps with SRE*, we will learn how operations can drive improvements such as automation in development.

3. **Collect feedback**: We learned in the previous chapter that DevOps embraces continuous testing as a quality measurement. This means that feedback is constantly collected. This is technical feedback and feedback on the functionality. Developers use this feedback to improve the next iteration or version. This is also a matter of the DevOps culture: feedback should not be seen as criticism or even a verdict. Feedback is really an instrument to improve quality throughout the project.

4. **Test**: In conjunction with collecting feedback, software is continuously tested. Does the code work properly and is it meeting the requirements? Testing is probably one of the most important things in DevOps projects. In *Chapter 3*, *Architecting for DevOps Quality*, we discussed testing strategies and different types of tests.

5. **Publish**: If the product has reached its final state, it's ready for go-live so that it can be used. Two topics that need attention in a launch are the user training and the period of after-care. Users will need to learn how to use the product and the software, and be prepared that as soon as products go into production and are actually used, issues still might arise. In after-care mode, teams can still pick up these issues fast.

    However, DevOps already takes care of this in itself. Issues create feedback that is looped back into the project, driving improvements. In practice, teams will create *fast lanes* to pick up issues in production with high priority. This might halt the further development of products and the development of new features. Exactly this is addressed in **site reliability engineering** (**SRE**), the main topic of the next chapter.

To summarize, the RAD process is shown in the following diagram:

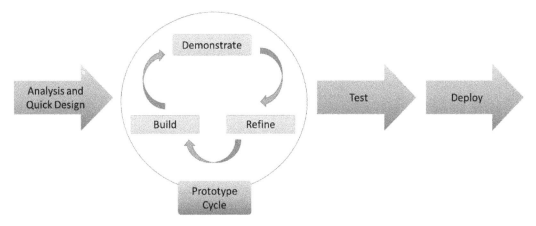

Figure 4.3 – RAD

In this section, we learned how to integrate software development in a DevOps project. Software needs infrastructure to run. In the next section, we will discuss scaling infrastructure.

Scaling infrastructure with DevOps

One of the key features in modern DevOps is the use of cloud technology. In this section, we will discuss why enterprises gain a lot of benefit by moving infrastructure in cloud platforms such as AWS and Azure. First, we will study the principles of scaling, since this is one of the major benefits of using cloud infrastructure. At the end of the section, we will also touch upon next-level scaling with containers, given the fact that in the coming years, there will be a big shift from VMs to containers.

In DevOps projects, developers use pipelines, as we have seen in the previous chapter. Code is pulled from a repository, changed, tested, and pushed to the next stage. Code follows a promotion path: from development to test, acceptance, and—eventually—production systems. Development and test systems might not always be needed; they simply have to be there whenever they are required in the process. If the work is done, then these systems might be suspended or even decommissioned. The benefit of the cloud is that enterprises don't pay for these systems if they're not in use, in contradiction with on-premises hardware that has been purchased as a one-off investment. So, the scaling up of development and test systems on demand is a huge advantage of cloud infrastructure.

Another important feature of modern DevOps is automation. A major benefit in using cloud infrastructure is the ability to have automatic scaling. However, architects and engineers might want to be a bit careful with automated scaling. It's true that enterprises pay for what they use in the cloud, whereas in the traditional way of working enterprises would need to buy physical machines whenever extra capacity was needed. The good side about that was that architects really needed to think about the required capacity.

In the cloud, there might not be a driver anymore to worry about capacity. Nothing could be further from the truth, though. In cases of peak demand and automated scaling without setting limits, the cloud bill will turn out to be a surprise. Enterprises, especially financial officers, should therefore not completely rely on scaling. In other words, architects will still have a responsibility to plan capacity.

Now, let's study the different varieties of scaling, as follows:

- **Vertical or scale-up**: Let's take a server as an example to explain this. The server has one processor, 2 **gigabytes** (**GB**) of memory, and 100 GB of disk storage. If we scale up, we add processors, memory, or disk storage to that server. We are adding resources to the same machine and increasing its capacity. That can be done as long as there's room in the server to add resources. We can imagine that this is easier in a coded, virtual world than with a physical machine where engineers really need to take out their screwdrivers and mount, for instance, extra memory cards in that server. The following diagram shows the principle of vertical scaling:

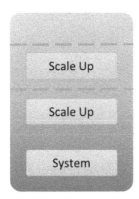

Figure 4.4 – Scale-up or vertical scaling

- **Horizontal or scale-out**: Now, we're adding more servers to our environment, instead of increasing resources within the server. This is very common in the public cloud, especially when we're using load balancers to handle traffic and spreading the workloads among the available servers. With automatic scaling, servers or pools of servers can be added automatically whenever the load is increasing and existing servers can't handle it anymore without degrading the performance. As soon as the load decreases, the environment is scaled down again. Load balancers—an example is **Elastic Load Balancing** (**ELB**) in AWs—make sure that the load is evenly spread across available resources.

However, keep in mind that applications need to be *scale-aware*. Some applications can't handle scaling at all, while some can scale out but can't be scaled down without impacting the availability of the application. The following diagram shows the principle of horizontal scaling:

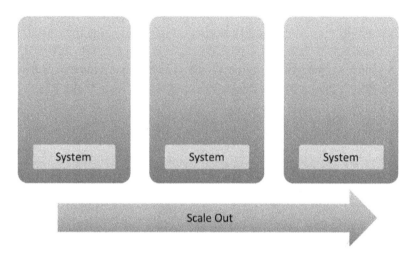

Figure 4.5 – Scale-out or horizontal scaling

- **Full or dynamic scaling**: This is a combination of vertical and horizontal scaling. As soon as limits are reached in scaling up, environments can be scaled out. In most cases, the scaling out is done by a cloning server. In Azure, we can use Azure Automation to do this. In AWS, we can copy the image of the server and then spin up a new machine with that image in **Elastic Compute Cloud** (**EC2**), using AWS Systems Manager Automation. Off course, there are a lot of third-party tools too that can help in automating scaling.

There are clear benefits of scaling in the cloud. We can have resources available on demand, scaling to the capacity that we need at a certain time. Since we're only paying for what we use, we could save money by scaling down resources if we don't need them anymore. For example, development and test systems might not be needed all the time and can be suspended when the work is done. Maybe teams can completely decommission these systems and simply spin up new ones as soon as the project requires it.

The best part is that DevOps teams can control this completely by themselves—they are no longer dependent on purchase departments that need to order hardware or require engineers to install it in the data center. It's all code, including the scale sets, and it can be fully integrated in the pipelines, ready at their disposal.

Scaling with containers

A major forthcoming change in IT infrastructure is moving from VMs to containers. The driver behind this is interoperability of systems between different platforms. Containers seem to be a very good solution to have software interoperable across platforms, with ultimate scalability. There are, however, a few things that an architect needs to consider. To start with, they must understand that containers also need infrastructure to land on. Containers do not run by themselves.

Containers are operated on compute clusters with a management layer that enables the sharing of resources and the scheduling of tasks to workloads that reside within the containers. Resources are compute clusters, a group of servers—commonly referred to as nodes—that host the containers. The management or orchestration layer makes sure that these nodes work as one unit to run containers and execute processes—the tasks—that are built inside the containers.

The cluster management tracks the usage of resources in the cluster such as memory, processing power, and storage, and next assigns containers to these resources so that the cluster nodes are utilized in an optimized way and applications run well.

In other words, scaling containers is not so much about the containers themselves, but more about scaling the underlying infrastructure. To make this a bit easier, Google invented the orchestration platform Kubernetes that takes care of cluster management. Kubernetes uses pods, enabling the sharing of data and application code among different containers, acting as one environment. Take the last sentence quite literally. Pods work with the **share fate** principle, meaning that if one container dies in the pod, all containers go with it.

The workflow in the following screenshot shows the basics of Kubernetes:

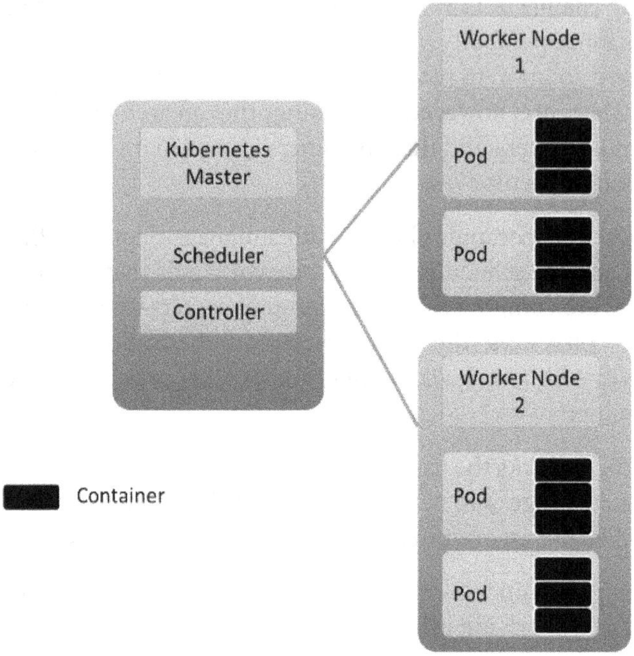

Figure 4.6 – High-level architecture of Kubernetes

The good news, though, is that pods can be replicated using replication controllers. Kubernetes polls whether the specified number of pods is running within the cluster nodes. If required, pods are replicated, making sure that the specified number of containers is running.

Containers are a good solution, but there are still some shortcomings. The most important one is that containers and clusters might be interoperable, but typically, networks and storage layers are not. In order to scale container solutions, we also need networks and storage layers to be *integrated*. For example, Azure Blob is a different beast from AWS **Simple Storage Service (S3)**, yet Kubernetes runs on both platforms using **Azure Kubernetes Services (AKS)** and **Elastic Kubernetes Services (EKS)** on AWS. There will be solutions to overcome this, but it's definitely something to take into account when planning container platforms.

Scaling DevOps in an enterprise environment

We've discussed the benefits of DevOps and what cloud adoption, automation, and an agile way of working could bring to an enterprise. The big question is: *How and where to start?* Opinions differ here, from a **big-bang approach** to step-by-step adoption.

Enterprises that have a lot of their IT muscles outsourced to different IT suppliers and that have been working for decades in a certain way are not easily changed. For one, there will be a lot of pushback from staff—remember that DevOps is also about changing a mindset or a culture. In this section, we're taking the approach of step-by-step adoption, or evolution instead of revolution.

Here are some recommendations:

- **Start small**: Don't start by implementing DevOps on large projects. Organize a small team and a simple project to learn and—even more important—to identify possible bottlenecks in the processes. What is possibly hindering the DevOps way of working? Do resources have the right skills, and is the team composed of the right resources? Does the team have the required tools? Are the requirements clear, even if it's a simple build? Are processes aligned with DevOps? Learn from the bottlenecks and improve in each step.

- **Start with the end in mind**: Know where you are going and what the end product will look like. Working in small iterations doesn't mean that the team will not need a clear picture of the end goal of the project. The same applies for implementing DevOps in an enterprise. From enterprise architecture, it must be clear what the strategy is for that enterprise: where will it be in 1, 3, or 5 years? Defining an enterprise roadmap can help in setting goals. An example is presented in the following diagram:

Figure 4.7 – Enterprise roadmap for adopting DevOps

The preceding screenshot shows three basic stages, outlined as follows:

- **Foundation**: The architect defines the target operating model, based on a reference architecture covering the applications, technology, security, services, and governance. At this stage, the cloud adoption is an important topic: the major cloud platforms, Azure, AWS, and Google Cloud, have **Cloud Adoption Frameworks (CAFs)** that will help in setting up the basics to operate systems in the cloud.

- **Adopt**: This stage is about adopting the foundation and the target operating model. The cloud environments are set up and the first—small—projects are initiated in DevOps mode. Concepts such as **Infrastructure as Code (IaC)** and RAD can be introduced. A way to do this is by installing a **Center of Excellence (CoE)** with **subject-matter experts (SMEs)** that can guide in the adoption of the model, including the cloud technology, use of DevOps tooling, and agile coaches to help implement the agile way of working. In the next section, *Scaling using a CoE*, we will discuss the setup of the CoE.

- **Expand**: We have referenced architecture, have defined a target operating model, and assigned a group of experts to take part in a CoE to help in adopting the new models and delivering the first projects. At this stage, the model can be expanded in the enterprise.

- **Make sure all steps are visible**: Transparency is key in DevOps. It applies to the way of working within the teams and the delivery process of products. Tools must enable full visibility as to what happens in the development and release chain, the CI/CD pipelines. Ideally, teams have a single-pane-of-glass view on events in the release chain: tools that collect real-time data from the pipelines and the systems. But also, team members need to know exactly what other members are doing, since DevOps is in essence mainly about close cooperation. Team members need to be able to track activities, anticipate, and—if needed—correct steps. The end goal is a better product.

- **Be ready for change—at all times**: This one seems obvious, but in DevOps nothing is set in stone. If something can be improved, teams should be motivated to adopt the change that enables this improvement. It applies to DevOps teams and their projects, but also to the enterprise as a whole. Even the biggest enterprises in the world every now and then have to hit the refresh button, to quote the book of Microsoft's **Chief Executive Officer (CEO)** Satya Nadella.

We've introduced the CoE. In the next section, we will elaborate on that.

Scaling using a CoE

A starting point could be a CoE. Again, it sounds like a big thing, but it doesn't have to be. On the contrary—a CoE can be a good entry point to start transforming a business. The CoE is a team that leads or supports employees and organizations in the adoption, migration, and operations of new technology, or even a new way of working. In short, the CoE can be the starting point of the digital transformation of an enterprise. Instead of trying to change the enterprise as a whole at once, we assign a team to guide this. The main goal of a CoE is to define and help implement best practices for implementing architecture, transforming and optimizing operations, and implementing governance.

The installation of a CoE should also be done in steps, starting with a CoE that defines the standards and policies. For that reason, the architect should be a member of the CoE. Next, the CoE defines the *guardrails*, ensuring the usage of best practices. Don't reinvent the wheel, but use what's out there and has been proven to work well in other enterprises. But there's a risk in that too. The risk is that the team is making it too big.

A commonly used framework to enroll an agile way of working is **SAFe**, the **Scaled Agile Framework**. It might include the implementation of the Spotify model, with the instalment of tribes and squads. Those are huge changes for any company, even if it's done in just one team. It will impact the whole enterprise, especially when IT is outsourced and resources from suppliers need to be involved in newly formed teams such as squads. Does the contract between the enterprise and the supplier even cater for the new way of working? Before you know it, we are implementing a world-conquering plan.

It doesn't mean that we can't use principles of SAFe, but we need to make sure that it *fits* and that it is adopted. The CoE can help in defining and controlling the adoption toll gates and suggest improvements. This type of CoE—still a small team—is referred to as prescriptive.

The next level of the CoE is the advisory level. At this stage, the CoE is formed as a (virtual) team of SMEs in different domains, actively helping DevOps teams in executing projects. The CoE guards the standards and policies, and controls and validates whether these are followed. From this point, the implementation of DevOps and agile is accelerated, breaking down the original organization silos. However, this is done step by step.

Starting with simple projects in small teams, it doesn't sound like DevOps and agile are really suitable to develop and run mission-critical environments. That's not the case. An enterprise might want to start with mission-critical, but if DevOps is scaled right and there's a clear plan on app modernization, we can also start managing critical environments in a DevOps way. The final section, *Managing mission-critical environments with DevOps,* of this chapter will explain more on this.

Managing mission-critical environments with DevOps

In this section, we will discuss DevOps for mission-critical environments and why it can be done to manage core applications. Let's first define **mission-critical**.

A very straightforward definition would be: any software that an enterprise needs to remain in business. If a mission-critical system were to fail, the enterprise would potentially lose a lot of money, either due to direct missed transactions or through things that are less tangible, such as reputation damage. These systems are identified through the process of **business impact analysis (BIA)**.

When we start with DevOps projects, the first thing that an architect does is gather the business and technical requirements. That would include the outcomes of the BIA process, which is typically done in cooperation with internal auditors and business stakeholders. From the BIA, critical systems or system components are identified that need to be restored very quickly in case these systems fail. This is a very cumbersome process that will cause of lot of discussion.

The enterprise architect will need to understand that stakeholders might have different views on what critical systems are. Financial systems in banks will be business-critical, but a car factory will not immediately lose business if the **Chief Financial Officer (CFO)** can't access financial reports for—let's say—an hour or so. Production at that factory, however, will stop immediately if the assembly robots fail.

Enterprises are still reluctant to host critical systems in public clouds because they think that they will lose control over the systems if these are not sitting in a privately owned on-premises data center that engineers can immediately enter in the case of an emergency. Yet, the public cloud might be the best place to host these systems. Because of the vast capacity that these platforms have, it's easy to have a copy of critical systems in different regions and zones. If one cloud data center fails, there's a second data center that can take over. Using cloud technology, this can be done with a minimal loss of data. Cloud technology offers tools to build more resilient environments.

Where does DevOps come in? In drafting the **business continuity plan (BCP)** for which the BIA is the input. There's no real technical reason why mission-critical systems can't be cloud-hosted and developed and managed through DevOps, but given the fact that these systems need to be highly resilient, there are a couple of things to consider—for instance, in planning and applying changes.

> **Note**
>
> There's a nuance to the statement that there's no real technical reason why mission-critical systems can't be hosted in the cloud. Latency can be an issue: the time that information needs to travel between systems. One other reason can be compliancy set by law and regulations. Some organizations are simply not allowed to host systems in a cloud data center that is not residing in the country or region where the organization itself is based. These are aspects that need to be taken into account too as part of the BIA.

An ongoing theme within DevOps is CI. That comes with changes, and changes have impact, also on business continuity. With critical systems, we have to make sure that the release process is designed in such way that business continuity is safeguarded. Quality assurance is, therefore, crucial.

First of all, test the code as soon as it's created. With critical systems, tests must be focused on ensuring that as code is pushed to production, the vital processes of the enterprise are not impacted or are only very limited, at an explicitly prior-accepted risk level. Next, be sure that there's a fallback, rollback, or restore mechanism.

How can teams be sure that what they're planning to release is *safe to go*? The answer to that one is a go-live run. Here's where the promotion path that we discussed in *Chapter 3, Architecting for DevOps Quality,* plays a crucial role. The go-live run is a real practice with the tested code on an acceptance system. That system should have exactly the same specifications as the production systems. Better: acceptance systems are an exact copy of production; they are production-like. The go-live run is done from the CI/CD pipelines, using the code as it's processed and pushed to acceptance. But it's not only about the code. Processes, security, failover to different systems, and restore procedures must be tested as well. DevOps tools need to be able to support this, as part of the BCP or framework.

This concludes the chapter. In the last section, we touched upon resilience and reliability. In *Chapter 5, Architecting Next-Level DevOps with SRE,* we go deeper into architecting for reliability with SRE.

Summary

This chapter covered a lot of ground. It's not easy to start with DevOps in large, traditional enterprises, but it is possible. In this chapter, we learned that we can start small and then slowly expand. Starting small doesn't mean that an enterprise doesn't need to have an end goal in mind: the enterprise architect has a key role in defining the target operating model and the way the enterprise will develop and operate products in the future. A CoE with SMEs can guide in this transformation.

There's a good chance that the company has legacy environments that will need to be transformed. We've discussed modern DevOps and using cloud and cloud-native technology. We also learned about different transformation strategies for applications and how we can develop new applications in DevOps mode using RAD.

In the last section, we also learned that even mission-critical systems can be developed and managed in a DevOps way, if we focus on resilience and reliability of these systems. SRE is a method to cover this. We will learn about architecture in SRE in the next chapter.

Questions

1. We are migrating an application from an on-premises system to Azure. The SQL database is migrated to Azure SQL as a PaaS solution. What do we call this migration strategy?

2. Name the Kubernetes services that Azure and AWS offer.

3. To assess business-critical systems, we need to analyze the requirements of these systems. What is the methodology for this?

Further reading

The Modern DevOps Manifesto (`https://medium.com/ibm-garage/the-modern-devops-manifesto-f06c82964722`) by Christopher Lazzaro and Andrea C. Crawford, 2020

5

Architecting Next-Level DevOps with SRE

In previous chapters, we discussed the ins and outs of DevOps. It's called DevOps for a reason, but in practice, the Dev is typically emphasized: creating agility by speeding up the development. **Site Reliability Engineering** (**SRE**) addresses Ops very strongly. How does Ops survive under the ever-increasing speed and number of products that Dev delivers? The answer is SRE teams, working with error budgets and toil.

After completing this chapter, you will have learned the basic principles of SRE and how you can help an enterprise adopt and implement them. You will have a good understanding of how to define **Key Performance Indicators** (**KPIs**) for SRE and what benefits these will bring to the organization.

In this chapter, we're going to cover the following main topics:

- Understanding the basic principles of SRE
- Assessing the enterprise for SRE readiness
- Architecting SRE using KPIs
- Implementing SRE
- Getting business value out of SRE

Understanding the basic principles of SRE

In this section, we will briefly introduce SRE, originally invented by Google to overcome the problem of operations completely being swamped by all the new developments that Google launched. There are a lot of definitions of SRE, but in this book, we'll use the definition used by Google itself: the thing that happens if you allow a software engineer to design operations.

Basically, Google addressed the gap between development and operations. Developers changed code because of demand, while operations tried to avoid services breaking because of these changes. In other words, there was always some sort of tension between dev and ops teams. We will talk about this more in this chapter.

Now, is SRE the next-level DevOps? The answer to that question is: SRE forms a bridge between Dev and Ops. A logical, next question, in that case, would be: is a bridge necessary? In the next section, we will learn that putting developers and operations in one team is simply not enough. There's a natural conflict of interest. So, an enterprise will need to do more to really get the benefits from working in DevOps mode. That is exactly what SRE does.

Key topics in SRE are reliability, scalability, availability, performance, efficiency, and response. These are integrated into seven principles – taken from the SRE Workbook – that are relevant to architecture:

- **Operations is a software problem**: This is the starting point of SRE. Software will change, yet operations need to remain stable so that services are not interrupted. It means that software needs to be resilient and tested intensively.

- **Work according to Service-Level Objectives (SLOs)**: Set clear targets for the service. What should really be the availability of an application? These are not solely IT-related objectives. Business requirements set the parameters in the first place. It means that projects have to work together closely with business stakeholders to define the objectives.

- **Work to minimize toil**: We will further discuss toil, but in principle, toil means just work. Moreover, toil is manual work that can be avoided through automation. Basically, the ambition of SRE is to get computers to do the work for you, as much as possible.

- **Automate**: After principle three, this one is logical. If you can automate it, do so.

- **Fail fast**: Failure is OK, as long as it's discovered at a very early stage. Fixing the issue in that early stage will have a much lower impact than when it's discovered and fixed at a later moment in the development and deployment cycle. Also, finding and fixing problems at an early stage will lead to lower costs.

- **Every team member is an owner**: This one requires a bit more explanation, especially in large enterprises where we typically have a matrix organization. Again, remember that a lot of enterprises work in sourcing models where specific suppliers are responsible for the infrastructure (the platform) and other suppliers and the enterprise itself for the applications (the products). In SRE, these boundaries don't exist. Both products and platform teams share the responsibility for the end product. Hence, products and platform teams need to have the same view on every component: application code, frontend systems, backend infrastructure, and security rules. They are all equal owners of the project.

> **Note**
>
> We will discuss this in more detail in *Chapter 6, Defining Operations in Architecture*, where we will learn about platform ops and product ops.

- **Use one toolset**: All teams use the same tools. You can have multiple SRE teams, but they all work with the same tools. The reasoning behind this is that an enterprise will have to spend too much time managing the different toolsets, instead of focusing on the project deliverables. And SRE is mainly about focus.

There's no way to summarize SRE in just a few paragraphs, or even one chapter. This section is merely a very short introduction. Still, SRE can bring a lot to an enterprise – if the enterprise is ready for that. SRE also means changing the way of working and with that, the organization. In the next section, we will learn more about that.

Assessing the enterprise for SRE readiness

In the previous section, we introduced SRE and discussed the basic principles, without the ambition of being comprehensive. Covering SRE as a whole would fill a book with well over 500 pages; we have merely given a quick overview of the most important parts. Now the question is: how do I know whether my company is ready for SRE? We will explore some criteria for SRE readiness in this section.

One of the common problems of companies implementing DevOps is that developers and operations are not really working together. They might sit in one team, but still there will be developers writing code and *throwing it over the fence* to operations when they think the code is done. The reason is that dev works with a different mindset than ops. Developers want to change. They get their assignments from business demand to improve or build new applications. Operators, on the other hand, don't want that change. Their main interest is to have stable systems that don't suffer from outages because of incidents or, indeed, changes. There's a conflict of interest, to begin with.

The question is how to bridge this conflict. SRE is the answer to that. However, SRE is a methodology that will only succeed if teams are ready to work with that methodology. So, one of the first things that we need to assess is culture. And yes: the enterprise architect does play a role in this. It's about processes and getting people to adopt these processes. Remember that an architecture is not only about the *what*, but also the *how*.

Redefining risk management

Developers will change code; operations need to be sure that the systems remain stable so that business is not halted. Changes might lead to downtime. If downtime is planned, then there's little to worry about. The key here is unplanned downtime. Hence, we need to focus on mitigating the risk of unforeseen outages due to changes. To avoid outages, systems need to be reliable and resilient. Since in DevOps iterations and changes are continuously deployed, the need to design for reliability becomes increasingly important. Architects need to design systems in such a way that they can handle changes without interrupting services.

Let's first agree on the definition of risk management. The basic rule is that risk equals probability times impact. Enterprises use risk management to determine the business value of implementing measures that limit either the probability and/or the impact – or, to put it in SRE terminology, risk management is used to determine the value of reliability engineering. Plus, it defines the level of investment to prevent, reduce, or transfer the risk.

Risk management is used to prioritize reliability measures in the product backlogs of SRE teams. That is done by following the risk matrix referred to as PRACT:

- **Prevent**: The risk is avoided completely.
- **Reduce**: The impact or likeliness that the risk occurs is reduced.
- **Accept**: The consequences of the risk are accepted.
- **Contingency**: Measures are planned and executed when the risk occurs.
- **Transfer**: The consequences of the risk are transferred, for instance, to an insurance company.

An example of a risk matrix is provided in the following template:

Risk ID	RISK DESCRIPTION	RISK ASSESSMENT		
		RISK SEVERITY	RISK LIKELIHOOD	RISK LEVEL
		ACCEPTABLE	IMPROBABLE	LOW
		TOLERABLE	POSSIBLE	MEDIUM
		UNDESIRABLE	PROBABLE	HIGH
		INTOLERABLE	PROBABLE	EXTREME

MITIGATIONS	REST RISK			
	RISK SEVERITY	RISK LIKELIHOOD	RISK LEVEL	ACCEPTABLE TO PROCEED?
	ACCEPTABLE	IMPROBABLE	LOW	YES
	ACCEPTABLE	IMPROBABLE	LOW	YES
	TOLERABLE	POSSIBLE	MEDIUM	YES
	UNDESIRABLE	PROBABLE	HIGH	NO

Figure 5.1 – Template for risk assessment

First, we need to identify and rate the risk. What is the risk, what are the chances that it will occur, and what will the impact be? That is shown in the top half of the figure. Then, we need to think about mitigations, actions that can or must be taken to prevent the risk or reduce the impact of the risk. When mitigating actions reduce the risk levels, then there will be a risk residue. Again, the team will need to assess what the impact of that residue will be and whether that's acceptable.

If the impact of failure is high, it might be worthwhile to look at a strategy that prevents the risk. This will drive the SLOs, or how good a system should be. If the availability is set to 99.99%, then the error budget is only 0.01%. This has consequences for the architecture of the system; after all, the risk rating allows for just 52 minutes of downtime per year. The architecture needs to cater for that, for instance, by having mirrored, hot standby systems that can take over as soon as primary systems fail.

But what about the code? There are two crucial elements in having resilient code:

- The source code needs to be stored securely in a repository with strong access and version control. Code change needs to be fully traceable.

- Automation of continuous tests to detect defects in code in every stage of the development and deployment. Tests and automation are likely the most important features architects will need to cover in DevOps, ensuring resilient and secure code.

Even if we as architects have done everything to prevent systems going down or software failing, we will encounter problems every now and then. An important rule within SRE is the blameless post-mortem. We will discuss that in the *Architecting SRE using KPIs* section.

Redefining governance

DevOps already assumes that teams are working in a highly autonomous way, meaning that they are responsible for the entire product from beginning to end. This requires different governance. In the previous chapter, we discussed the center of excellence as an organization to guide and support the DevOps teams. The center defines the overall enterprise roadmaps and frameworks that provide the guardrails to develop and manage systems.

Now, DevOps is still about dev and ops, virtually still divided. SRE teams don't have that division. SRE teams are for that reason different from DevOps teams. SRE teams are cross-domain, meaning that they focus on monitoring the systems, logging, and processing events and automation. They help in developing and implementing automation, but also advise and guide in doing releases in the DevOps process. SRE engineers are able to help in defining system architecture, but can also assist in advancing the enterprise architecture by advising in best practices and selecting the right tools.

There are three ways of setting up SRE teams:

- **Team with dedicated SRE engineers**: These teams are separated from the DevOps teams, but support the DevOps teams. A big advantage is that a lot of teams and different projects are supported at the same time, with the same vision, tools, and processes, improving the overall quality of the different projects.

- **Embedded model**: SRE engineers are embedded in the DevOps teams. The advantage of this approach is that SRE engineers can focus on specific issues within projects they are assigned to.

- **Distributed SRE model**: In this model, the SRE team works more as a center of excellence with specialists that can be consulted to solve issues.

A typical way to position SRE is presented in the following figure:

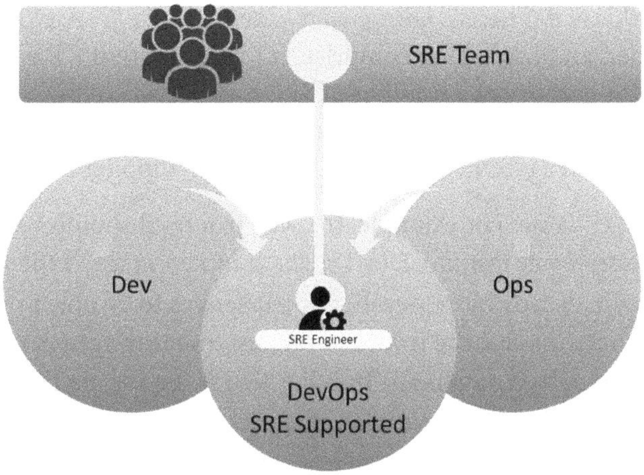

Figure 5.2 – Position of SRE in DevOps

One thing must be absolutely clear, and that is that SRE requires a change of culture. SRE focuses on improving operations, yet also facilitating development and releases at high velocity. It often means that SRE specialists require a high level of standardization of technology and processes. If development and operations are standardized, then it's also easier to automate processes. By doing that, SRE drives down the probability that risks materialize. As a result, engineers are freed up to pick up other tasks instead of having to spend a lot of time solving problems. That's the key takeaway of SRE.

The next question is how an organization can implement SRE. That starts with defining KPIs. In the next section, we will study the most important KPIs in SRE.

Architecting SRE using KPIs

Before we dive into the definition of KPIs, we need to get back to the basic principles of SRE. SRE teams focus on reliability, scalability, availability, performance, efficiency, and response. These are all measurable items, so we can transform them into KPIs. In this section, we will learn how to do that using SLOs, **Service-Level Indicators (SLIs)**, and the error budget.

The main KPIs that we use in SRE are as follows:

- **SLOs**: In SRE, this is defined as *how good a system should be*. An SLO is much more precise than an SLA, which comprises a lot of different KPIs. You could also state that the SLA comprises a number of SLOs. However, an SLO is an agreement between the developers in the SRE team and the product owner of the service, whereas an SLA is an agreement between the service supplier and the end user.

  The SLO is a target value. For example, the web frontend should be able to handle hundreds of requests per minute. Don't make it too complex at the start. By setting this SLO, the team already has a number of challenges to be able to meet this target, since it will not only involve the frontend but also the throughput on, for instance, the network and involved databases. In other words, by setting this one target, architects and developers will have a lot of work to do to reach that target.

- **SLIs**: SLOs are measured by SLIs. In SRE, there are a couple of indicators that are really important: request latency, system throughput, availability, and the error rate. These are the key SLIs, measuring how good a system really is. Request latency measures the time before a system returns a response. System throughput is the number of requests per second or minute. Availability is the amount of time a system is usable to the end user. The error rate is the percentage of the total number of requests and the number of requests that are successfully returned.

- **Error budget**: This is probably the most important term in SRE. The SLO also defines the error budget. The budget starts at 100 and is calculated by deducting the SLO. For example, if we have an SLO that says that the availability of a system is 99.9%, then the error budget is *100 − 99.9 = -0,1*. This is the room that SRE teams have to apply changes without impacting the SLO. It forces developers in the SRE team to either limit the number of changes and releases or to test and automate as much as possible to avoid disruption of the system and overspending the error budget.

To understand the concept of error budget in SRE, it's important to understand how SRE treats the availability of systems. It's not simply a matter of deducting the downtime to get to the availability of systems. SRE takes failed requests into account. A failed request can be because a system doesn't respond or has a slow response. Detecting failed requests determines the availability and thus whether the error budget is exceeded or not. Important parameters are as follows:

- **TTD**: The **time to detect** an issue in software or a system.
- **TTR**: The **time to resolve** or repair the issue.
- **Frequency/year**: The frequency of error per year.

- **Users**: The number of users that are impacted by the error.

- **Bad/year**: The number of minutes per year that a system is not usable, or the *bad minutes* per year.

Working with the error budget is shown in the following workflow:

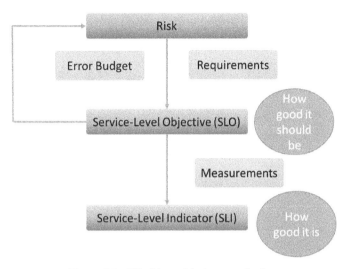

Figure 5.3 – Working with the error budget

Risk should be perceived as business risk: something endangering the business of the enterprise. Recognizing that risk leads to requirements of systems and software. These requirements are translated into SLOs, defining how good a system should be. SLOs are measured by indicators, telling exactly how good the system really is. If SLOs are not met, they will trigger the risk to materialize. The chance that systems might fail and with that SLOs are not met is the error budget. The error budget – typically when the budget is exceeded – will lead to adjusted requirements and improved systems.

Despite all the work we put into reliability, enterprises will be confronted with issues and, as a consequence of that, outages. A key element in SRE is the blameless post-mortem that can be executed on different levels. We can have post-mortems whenever an incident occurs or after a project has been completed. The blameless post-mortem is really all about culture: it investigates incidents without blaming. SRE teams simply assume that all involved team members have done their very best to avoid the incident.

The teams evaluate the issue and come up with advice to avoid the issue occurring again. This can be improvements to the process or the use of tools. Also, advice might be about people, for instance, to get people trained in specific areas. This is not to blame, but always to improve.

If we put it all together, SRE is a truly holistic model. It's about processes, tools, and people. The following diagram shows a holistic view of SRE and how it merges with DevOps:

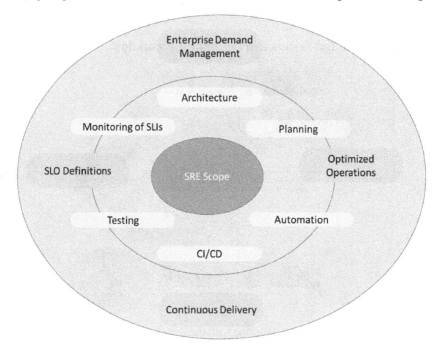

Figure 5.4 – Holistic view of SRE

In this section, we introduced the main aspects of SRE. The top question and challenge for most enterprises is: where do we start? We will discuss that in the next section.

Implementing SRE

So far, we have learned what SRE is and what the key elements are. In this section, we will learn how to start with SRE, but like DevOps, the advice is to start small. Then there are two major steps that will help you to implement SRE in a controlled way:

- **Agree on the standards and practices**: This can be for just one SRE team or for the entire enterprise if the ambition reaches that level. In some workbooks this is called **kitchen sink**, meaning that everything is SRE. This can be a viable approach for companies with a limited set of applications, but for enterprises, it might be wiser to work with an SRE team charter.

Let's work with a very common example that we will also use in the next chapters. Enterprises usually have product teams working on applications and a platform team that is responsible for the infrastructure. It's good practice to have an SRE team bridging between one product team and the platform team, setting out standards and practices for this particular domain. The product team will focus on the delivery of the application, obviously working closely together with the platform team. The SRE team can guide this and set standards for the reliability of the end product. It means that SRE teams have to cover multiple domains, as shown in the following figure:

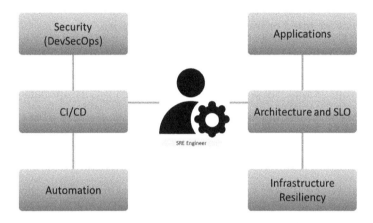

Figure 5.5 – Domains of SRE

- **Agree on the scope of services**: Certainly, at the beginning of the SRE journey, SRE teams can't do everything. Therefore, we need to agree on the scope of the SRE team. Will they only do consulting or will they be actively involved in the DevOps projects? Or will they only be involved in the automation of DevOps? Some companies have specific SRE teams for tooling, a tools-only SRE team that is only concerned with implementing automation tools, overarching DevOps. The final step in SRE is *on call*, whenever issues arise during operations and exceed the error budget. Then, SRE engineers are called in to investigate the issue, guide in the post-mortem, and help implement new solutions that ensure reliability.

Since SRE is a Google invention, Google has published extensive guidelines to implement SRE step by step. Google identifies three entry acceptance criteria to start with SRE:

- **SLOs** have been defined and agreed with business owners.
- **Blameless post-mortems** can be executed.
- The enterprise has a process in place to manage production issues. This can be the standard **incident management** process as defined in IT service management frameworks such as ITIL.

Where does SRE start and how do teams begin their work?

1. SRE specialists are hired or trained within the enterprise.
2. Release processes are documented and evaluated by SRE specialists.
3. Operational processes are documented, including runbooks for releases and handover to operations. These processes are evaluated by SRE specialists.
4. SLOs have been defined and agreed upon.
5. SRE teams are mandated to adapt and implement processes that reduce toil. Collaboration with developers and operations resulting in a buy-in is a prerequisite.

The latter is crucial, but even more important is that SRE teams do this without *blaming* as discussed in the previous section about the blameless post-mortem. Every single assessment of procedures, processes, and the root cause of failing systems should be without pointing fingers, but only focusing on improvements. SRE teams can do this for all new and existing processes:

- SLO and error budget review
- Incident reviews
- Test and runbook reviews
- Security audits

However, before the team can really start, the organization needs to have clear priorities. As we have learned, SRE will trigger a change of culture and the whole organization will have to support the implementation of SRE. Apart from the model that we use to implement SRE teams – dedicated, embedded, or distributed – the guidelines and guardrails of SRE will impact the entire enterprise. That means that the enterprise must have a strategy.

Typically, enterprises do not start from a *greenfield* situation. There are existing products, projects, and processes that can't be changed overnight. Google recognized this when it introduced the methodology. Enterprises are usually *brownfield* that enter a transformation. In that case, they need to think of priorities and how to start the transformation. In 2013, Google's SRE engineer Mickey Dickerson came up with the Hierarchy of Reliability. The model is as follows:

Figure 5.6 – Dickerson's Hierarchy of Reliability

The idea of the pyramid is that the items at the bottom are the basics that need to be implemented first; they form the foundation. From there the transformation progresses to more advanced items, such as the release chain in the development and deployment of new products at the very top of the pyramid.

We have identified our goals and objectives, assigned teams, and agreed on the priorities in the implementation. But will SRE actually bring benefits to the enterprise? We will answer that question in the final section of this chapter.

Getting business value out of SRE

As we have learned in the previous section, enterprises don't implement SRE in just a few days. It takes time and stamina to get it right. But is it worth it? Obviously, the answer is yes. SRE will allow for great business value. In this final section, we will explain how.

Today's enterprises are continuously transforming. It puts a lot of pressure on operations that on one side have to keep up with developments and on the other side have to keep systems stable and reliable. Without true collaboration between developers and operations, that's virtually impossible. SRE addresses this challenge. SRE recognizes that putting dev and ops together in one room doesn't solve the problem. SRE creates a solution that reduces operational issues by helping developers to build reliable systems. Key components are as follows:

- **Standardization**: Standardize processes and tools.
- **Automation**: Automation leads to consistency, but automation also enables scaling. This requires a very well-thought-out architecture. Automation is about doing something once and then letting automation take care of the rest. Without automation, operations would simply be drowned by manual tasks.
- **Eliminate toil**: Toil is manual work, repetitive, and can be automated. But toil is also work that doesn't add value to the product: it's interruptive and slows down the development of services that do add value.
- **Simplicity**: Software needs to be simple as a prerequisite to a stable, reliable system. Code needs to be simple and clean and APIs as minimal as possible. SRE lives by the golden rule of less is more.

By doing this, the enterprise will benefit from high-velocity developments, yet spending fewer resources in fixing issues. Time and resources can then be invested into further improvements. So, enterprises can gain a lot from adopting SRE. Because SRE involves a very systematic way to build, manage, and review systems, enterprises can trust reliable services. Repetitive tasks are taken over by standardization and automation. The business gains on multiple sides here:

- Reliable services will gain trust with customers and likely more revenue.
- Because of automation, manual tasks are reduced. This will drive the costs of operations down. The very first thing that SRE engineers will do is automate repetitive tasks.
- Because of standardization, systems will become more reliable and stable, causing fewer issues and outages. Analyzing issues and solving problems require resources and are therefore very costly. They don't add value to the business, but they need to be addressed in order to not stop the business.
- Cost savings might be invested in the development of new features and new products. In other words: SRE will be a driver for innovation. SRE will bring a lot of benefits to an enterprise, but implementation requires dedication and the willingness to adopt a different culture. As with anything in Agile and DevOps, the recommendation is to start small and then scale up throughout the enterprise. Learn from mistakes, optimize, and continuously improve.

Summary

This chapter covered the basics of SRE. The original workbook contains well over 500 pages, so it's almost impossible to summarize the methodology in just a few pages. Yet, after completing this chapter you will have a good understanding of the founding principles of SRE, starting with the definition of SLOs to set requirements on how good a system should be. Subsequently, we measure the SLOs with indicators that tell us how good the system really is. We learned that by working with risk management, error budgets, and blameless post-mortems, SRE engineers can help DevOps teams to improve systems and make them more reliable.

The conclusion of the chapter was that SRE is not very easy to implement in an enterprise. We discussed the first steps of the implementation and learned that if done right, SRE will lead to benefits. Businesses will gain from SRE because a lot of manual work can be reduced, creating room to improve products or develop new ones.

This concludes the first part of this book. In the next part, we will take the next step and learn how modern technologies can help the enterprise and further optimize operations. One of the promising new technologies is artificial intelligence and with that, we introduce AIOps in the second part of this book.

Questions

1. What is the term that SRE uses to label repetitive, manual work that should be reduced?

2. What do the terms TTD and TTR mean?

3. What do we do when we transfer risk?

Further reading

- *Multi-Cloud Architecture and Governance*, by Jeroen Mulder, Packt Publishing, 2020

- *Practical Site Reliability Engineering*, by Pethuru Raj Chelliah, Shreyash Naithani, and Shailender Singh, Packt Publishing, 2018

- Do you have an SRE team yet? How to start and assess your journey: `https://cloud.google.com/blog/products/devops-sre/how-to-start-and-assess-your-sre-journey`

Section 2: Creating the Shift Left with AIOps

In this part, you will learn about the role and mandate of ops in an enterprise and how this is embedded into the architecture. You will also learn that the role of ops is changing. In DevOps, ops are linked to developers. The aim of DevOps is to increase the velocity of development. As a consequence, ops will need to be more efficient, spending less time running operations. This is what AIOps brings to the table, automation and intelligence to discover abnormalities much faster and even solve them with auto-remediation. How do architects design systems that are ready for AIOps and maybe even the shift left to NoOps?

The following chapters will be covered under this section:

6
Defining Operations in Architecture

The role of operations is changing in enterprises that adopt DevOps. The number one task of operations is to keep services up and running, but in the new digital operating model, a lot of these tasks can and will be automated. Before we start thinking about automating operations, we need to have a clear view of roles, mandates, activities, and domains of operations in the enterprise. This chapter is the guardrail to capturing ops in the enterprise architecture. We will learn how to design architecture for operations and define the digital operating model.

After completing this chapter, you will have learned what the roles and responsibilities are of operations and how this can be addressed in the architecture. We will also see how operations is changing and impacted by the cloud, cloud-native services, and event-driven architectures that use microservices. We will design a digital operating model while making a distinction between platform and product operations. Finally, we will discuss how to elevate the enterprise to continuous operations.

In this chapter, we're going to cover the following main topics:

- Understanding operations management
- Defining operations in an enterprise architecture
- Defining the digital operating demarcation model

- Understanding ops in an event-driven architecture
- Planning operations with a maturity model

Understanding operations management

Before we can start defining operations management in an enterprise architecture, including the roles and demarcations between these roles, we need to understand what IT operations include. In this section, we will discuss the definition of IT operations and **IT Operations Management** (**ITOM**).

First, why is this important? DevOps has a tendency to focus on development: exploring and building new features and new products. In discussing release management and CI/CD, there's also focus on the development and deployment process. But operations is just as important as development, and the role of IT operations is changing. Not only because of DevOps, but also because of the digital transformation that a lot of enterprises are going through. We will learn more about this in this section.

As a short definition, we can say that IT operations includes all the processes that support hardware and software that are used by the enterprise to fulfill customers' services. So, IT operations is responsible for the functionality of end user devices such as laptops, but also for products that deliver services to customers of the enterprise. A simple example is a website where customers can order goods, including its infrastructure (web server) and application code (frontend application and database). IT operations also have a major task in assuring the quality of IT assets.

The following processes are important in ITOM:

- **Monitoring**: IT operations are the eyes of IT, so robust, resilient monitoring systems are crucial. There's not really a *one size fits all* approach when it comes to monitoring. Operations will work with different systems to control the infrastructure, the applications, interfaces, backup jobs, and many other components. The challenge is to get one overview out of these systems so that system states, failures, and potential issues can be correlated. For example, a web service might be not responding because the database is not available. End-to-end monitoring is a term that is often used in IT, and it means that monitoring systems simulate transactions throughout the chain of IT systems.

- **Incident management**: Any event that disrupts the normal operation of systems is an incident. It's up to operations to identify the incident (using monitoring) and solve it as soon as possible. This can be done by applying a temporary workaround to get systems back to normal, but it's best to fix the issue and make sure it doesn't happen again. That's what problem management is all about.

> **Tip**
>
> Being involved in operations for quite a number of years as an architect, there's one important lesson that every engineer and architect should learn well: there's nothing as permanent as a temporary solution to quickly solve an issue. It might be the fastest way to get systems back up and running, but in the long run, these quick fixes will cause new issues. One reason for this is that quick fixes are often not very well documented and after a while, no one knows how they were applied.

- **Problem management**: Here, incidents are analyzed in more depth. Also, trends in incidents are explored. Operations will need to align with engineers and architects to come up with solutions to prevent incidents from occurring again.

- **Change control and release management**: Operations will be confronted with changes to systems and in DevOps, these changes will occur quite often. However, it's the responsibility of operations to ensure that systems keep running without major, unplanned interruptions. This is the change control process. Part of this process might be running a final backup before the change is executed so that system states and data are secured. Developers and operations need to be fully aligned in applying changes. This is done in the process of release management.

In the more traditional way of working, operations will simply receive the new release and then decide when this is deployed so that existing services can continue without any disruptions. In DevOps, the process works differently. Here, the team has a shared responsibility in deploying the new release.

In short, operations must keep the services up and running under all circumstances. DevOps and digital transformation have a significant impact on operations. Let's review some trends:

- **Cloud and cloud native**: This may sound obvious, but cloud and cloud-native technologies have an enormous impact on IT operations. The contradiction is that a lot of these technologies aim to reduce the complexity in the IT landscape, but on the other hand, they add complexity with cloud-born assets. The IT landscape is turning more and more into an API ecosystem, where operations have to manage not only virtual machines, applications, and network connections, but also **Application Programmable Interfaces** (**APIs**) that connect the PaaS and SaaS solutions to the IT landscape of the enterprise.

- **Data center decommissioning**: With enterprises moving IT systems to cloud platforms, it's logical to assume that the private owned data centers are being emptied and decommissioned. Here, traditional operations work, including datacenter management, is not disappearing, but being transformed into managing the virtual data center in cloud platforms. Once more, everything is turning into code, whereas in traditional data centers, operators would still have physical machines to look after.

- **Faster networks**: There's less of a need to have systems near the enterprise itself. Systems can be hosted in global cloud platforms. High-speed network connections solve problems with latency, so there's virtually no limit to where systems can *land*. In the near future, we will see faster networks coming to the market: it's a requirement for a lot of new services that are being developed. Think of real-time data analytics, quantum computing or simulation, and technologies that rely on high-speed connections to, for instance, transmit images across the world. Networks are the foundation for everything, and operations will need to focus more on the resilience, agility, and performance of these networks.

- **Globalization**: Global cloud providers such as Azure, AWS, Google Cloud, and Alibaba Cloud have data centers around the globe. There's a big benefit in being able to have disaster recovery enabled in a completely different region, thus increasing the resiliency of systems. However, there's also a lot to consider, such as if enterprises are bound by legal regulations to have data within the borders of the country or region where they reside or deliver services. Globalization of IT has pros, but also cons.

- **Shift to left-left**: The idea of knowledge sharing in IT has gained a lot of momentum over the past few years. Shift left also means that IT systems cater for *self-help*: systems are designed in such a way that users can easily find ways to overcome issues. With shift left-left, an element is added to this: systems are learning from users and adapting software so that the issue doesn't occur again. This is where the final trend will play a huge role: **artificial intelligence (AI)** and **machine learning (ML)**.

- **AI and ML**: Last but not least, one of the major trends that will change the operations domain is AI and ML. This includes self-learning, self-healing systems, and systems that can even predict certain behavior and act accordingly to that behavior. In operations. we will see diagnostic systems that learn from issues and either fix these or give advice on how to deal with them. These systems are also capable of correlating events: if a system in the chain fails, AI will know how it will impact other systems in that chain and take mitigating actions, for instance, taking a snapshot of the system's state so that the root cause can be identified faster and more accurately.

At this point, you should know that operations will not become easier and that IT won't become less complex. This complexity will shift and transform. The good news is that these new technologies will allow you to predict system behavior better, design for more resiliency, and reduce risks further by finding the root causes of issues faster and taking swift, accurate corrective actions.

How does this fit into our architecture? We will discuss this in the next section.

Defining operations in an enterprise architecture

IT operations is not a part of the enterprise architecture, meaning that an enterprise architecture doesn't define how an enterprise must perform operations. At best, the operations architecture is part of the technical architecture. In this section, we will elaborate on the components of the enterprise architecture and then study the specifics of the operations architecture.

The enterprise architecture includes the following architectural components:

- **Business architecture**: This covers the business functions and the organization of the business, including its operations. From the business architecture, it should be clear how products and services are delivered, as well as what processes are followed for designing, building, and running them. The business architecture starts with the product strategy, which includes describing the products and services that the enterprise delivers.

- **Governance architecture**: This defines who's responsible for fulfilling processes. This is the blueprint for how the enterprise operates, including the tactical plans for IT that state how processes are implemented, operated, and monitored.

  Ops are key components in both the business and governance architecture. Ops is responsible for delivering products and services in a stable manner throughout the whole delivery chain. It's important to realize that there is a difference between operational service management and operations management, with the latter often referred to as ITOM. Service management comprises the typical ITSM or ITIL processes that we discussed in the first section about *Understanding operations management*, such as incident and problem management. ITOM is more IT technology oriented and focuses on the operations of applications and infrastructure. The enterprise architecture also addresses this.

- **Data architecture**: This describes how data is used; that is, why and by who or what process to fulfill a service. Operations have a role in making and keeping data available according to the data principles of an enterprise. These principles typically involve data privacy and compliancy rules. Due to this, operations need to work closely with security and data privacy officers.

- **Applications architecture**: This describes the build and usage of applications. Again, operations play a key role in keeping the applications running, including databases and middleware. It's highly recommended to involve operations in the very first stage of application development, to make sure that applications can really be managed by operations. Think of applying new cloud-native technologies to applications: an enterprise needs to be sure that operations have the skills to operate these technologies.

- **Technology architecture**: Finally, this describes all the technical components that are used in the enterprise. It should include products such as hardware and software, standards and principles, services, and policies.

The following diagram shows the levels in the enterprise architecture:

Figure 6.1 – Components of the enterprise architecture

The operations architecture can be part of the technology architecture, but at a more detailed level. It includes the following components at a minimum:

- Production scheduling/monitoring

- System monitoring

- Performance monitoring

- Network monitoring

- Event management (incidents, problems, changes)

Two items deserve special attention:

- **Service-level agreements (SLAs)**: The contract between a supplier and a customer that describes exactly what conditions a service is provided with, typically using **key performance indicators** (**KPIs**). Operations need to deliver the service according to these KPIs.

- **Operating-level agreements (OLAs)**: Part of the SLA can be OLAs, defining the interdependent relationships between components, and composing the service that is covered by the SLA. For example, the SLA might describe a web application that needs to be available 99.9% percent of the time. The application itself might be depending on database services that are not part of the application chain and are operated by a different ops group. The OLA will address this interdependency.

In this section, we concluded that IT is becoming more complex. Enterprises expect more from operations to keep systems stable, but also to keep up with new developments. To allow ops to fulfill these increasingly demanding tasks, they need an adjusted operating model that deals with this digital transformation. For that, they also need the right tools. In the next section, we will discuss this new, digital operating model before we step into the world of tools with AIOps.

Defining the digital operating demarcation model

The role and position of operations is changing; we saw that in the first section of this chapter. Besides new and evolving technology impacting operations, the most important reason for this is the shift from projects to product-oriented continuous delivery.

What do we mean by this? Most enterprises used to work in projects, typically waterfall type projects. There's a specific end date and the whole project is set out in a timeline with milestones. In DevOps, the focus is on the product: it starts with a Minimal Viable Product, and then the teams keep improving it in short sprints of 2 or 3 weeks.

At the end of the sprint, the product and the deliverables are reviewed. The developers and operations collaborate with these teams. In the traditional model, operations would get a final product and then decide how and when it would be released. The new operating model has to be more agile, adaptive, and embedded in DevOps. In this section, we will look at that new operating model in more detail.

It's important to understand what the role of dev is and what the role of ops is in a digital operating model:

- Dev designs and deploys.
- Ops fulfills and manages, including detection and correction. However, in DevOps, detection and correction is looping back into dev.

To create agility, we need to set **demarcation** in the different ops tasks. We will have operations focusing on the products and operations focusing on the platform. This is a requirement if we wish to provide product-oriented delivery. In a digital operating model based on agile and DevOps, we have businesses working together with product teams to define the necessary products and their features. Product ops engineers will support the delivery of such products, while the platform engineers will make sure that the platform – the infrastructure – is ready to land the products and provide an enduring stable service.

The following diagram shows a model with demarcation layers:

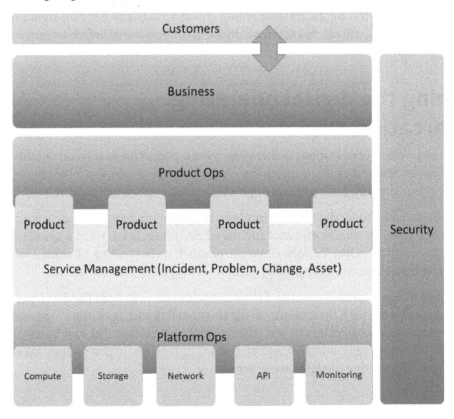

Figure 6.2 – Layered demarcation model with product ops and platform ops

In the next section, we will explain the model further:

- **Platform Ops**: The platform is the landing zone – the foundation. It needs to be there, and it needs to be stable. Product teams should not have to worry about the availability of the platform. In most digital operating models, the platform is usually a (public) cloud platform.

  The platform team that manages this is decoupled from the product teams. Platform operations include updates, upgrades, and optimization and also deploys new infrastructure features. These are planned in close cooperation with the product teams to avoid disruptions to services.

  Roles that are advised in platform operations are (cloud) platform manager, (cloud) platform architect, and platform engineers. There might be a great need to have an API integrations specialist in this ops team as well, since platforms are tending to become more like API ecosystems. These APIs, such as SaaS and PaaS solutions, need to be integrated with the platform.

- **Product Ops**: This is the layer where the product is designed, build, deployed, and managed in DevOps mode. Simply put, this is the layer where the product is defined as a basic idea and through iterations, all of which are eventually brought under product ops control. In this team, we will need platform engineers that bridge the requirements of the product to the platform. These platform engineers need to be trained and skilled in Infrastructure as Code, Configuration as Code, automation, and API programming. They will liaise with platform operations.

  Ops roles that are advised in this layer and the product team are infrastructure engineers and testers, developers, and specific domain architects who are involved in designing the product.

  One of the pitfalls of introducing agile and DevOps is that enterprises might step away from the traditional IT service management processes. However, the operational processes that we briefly discussed in the first section about *Understanding operations management* are still very valid. This is why, in a digital operating model, we need roles and responsibilities to address these processes:

  - (Major) Incident manager

  - Problem manager

  - Change manager

  - Asset manager

We can do this using a RACI matrix. RACI stands for **Responsible, Accountable, Consulting, and Informed**. The following table represents a simple example of a RACI for service management processes:

Role	Architect	Incident Manager	Change Manager	Engineer	IT Manager
Task					
Incident Management - Investigate incident	C	A		R	I
Incident Managment - Design solution	R	A	I	C	I
Incident Management - Implement solution	C	A	I	R	C
Coordinate change	C		A		I

Figure 6.3 – Example of a RACI matrix for service management processes

The placing of the **R**, **A**, **C**, and **I** are debatable, which is fine – as long as it is perfectly clear to all involved who's responsible for what.

- **Business**: This is the top level and is where the strategy is laid out and the requirements for the products are defined. In the new digital operating model, new roles are added to this layer, such as customer journey analysts and designers. The whole idea behind the model is that the enterprise caters for more agility and gets new products to the market faster. The business will need to know what customers want and how they experience the products to make them successful.

  This is also important for operations. Remember that dev and ops have to be aligned from the beginning to the end. Ops needs to be involved in the customer's journeys too. A specific ops role in this case is the service manager role: they need to be aware of what's coming and how this can be adopted in the new services, all while ensuring the existing services are not disrupted.

There's one important layer that's missing, but this is actually not a separate layer. Security and security management is an overarching layer and should be embedded in every other layer.

So, we have engineers and architects on all the layers, closely collaborating. But it should be clear that they need some sort of framework where they can work. That is the enterprise architecture. The enterprise architect sits at the very top of the model, closely working together with the customer journey designers, the domain architects in development, and the architects in operations.

IT4IT, by The Open Group, addresses this new model and suggests a way forward for enterprises. IT4IT marks four stages or *value streams* for products. These value streams are very accurate since IT creates value – a product being developing from an idea and becoming a real service:

1. **Plan**: Strategy to portfolio

2. **Build**: Requirement to deploy

3. **Deliver**: Requirement to fulfill

4. **Run**: Detect to correct

These value streams are shown in the following diagram:

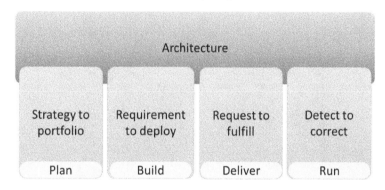

Figure 6.4 – IT4IT value streams

The deliver and run value streams are the ops streams. Request to fulfill includes cataloging, fulfilling, and managing the usage of the service. Run is about anticipating and resolving issues. Ops can help with these tasks by using an event-driven architecture. We will discuss this in the next section.

Understanding ops in an event-driven architecture

Let's review the most important task of ops once more: keeping services up and running. To enable this, we have defined a number of processes that help manage systems. Incident and problem management are key processes; that is, in IT4IT terms, detect to correct. The issue is that incident management is almost by default reactive: an issue is detected and actions are triggered to find and fix the issue. In the next phase, typically in problem management, a deeper analysis is done, where solutions are designed to prevent the issue from happening again.

Event management is a component of operations. The challenge in a digital operating model is to orchestrate and automate these events across different IT systems and even platforms. The event-driven architecture addresses this and is actually the starting point of AIOps. We will discuss this in more detail in *Chapter 8, Architecting AIOps*.

The event-driven architecture was originally meant to help design applications so that they react to events. An event is defined as a change of state that triggers a reaction. The following diagram provides an example of a build in Archimate:

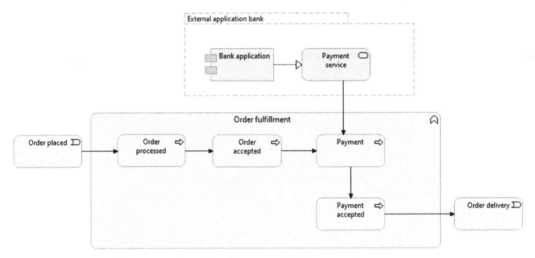

Figure 6.5 – Simple Archimate model for processing payments

Here, we have a business event: **Order placed**. This triggers a business function, which in this example is called **Order fulfillment**. Part of this function is **Payment**, which triggers an external system at the bank that delivers the payment service. As soon as the payment is accepted, the order leaves the function to the next business event, which is the **Order delivery**.

> **Tip**
>
> Archimate is a recommended methodology for designing functions and mapping these to processes that need to be fulfilled by IT components. Archimate uses viewpoints to model business processes to applications and technology. A common viewpoint is the product viewpoint, which shows the value that a product delivers to customers. The architect can use Archimate models to define the composition of the product, including the different services and interfaces between the services. Archi is a free tool that can be used to design Archimate models. Archi can be downloaded from `https://www.archimatetool.com/download/`.

In this – very simple – example, the customer pays their placed order. The order status now changes to paid and triggers a process that pushes the order to delivery. In fact, the payment process itself will also trigger other new events, such as a connection that needs to be established with a bank or payment service. Within the bank, there will be a trigger that checks the back account of the buyer and sends a message to the order system with a status of OK or not OK.

In short, a change of state will trigger a lot of new events. These events do not have to occur in the same system, as we have seen in this example. Events can drive decisions in other systems that are connected through APIs. This is the foundation of microservices and the **service-oriented architecture (SOA)**. Microservices are contained, independent operating services that communicate with each other using APIs. These services are managed by self-contained teams that develop, build, and run these services. So, it's fair to say that the event-driven architecture using microservices is well supported by DevOps with teams that are end-to-end responsible for delivering that particular service.

To enable this type of architecture, we need services to be defined as separate functions that are reusable, scalable, and interoperable. These are the corner stones of SOA and in fact, a lot of cloud technologies are derived from SOA principles. PaaS and SaaS solutions are defined as SOA, which means that they can be reused and shared in different environments, are scalable, and are interoperable between different platforms.

How does this all impact operations? To put it very simply: they have to keep services – microservices, to be precise – up and running across a lot of different applications, systems, and platforms. Ops must deal with decentralized continuous delivery by using interfaces that connect independent deployed services. The traditional way of monitoring will not be sufficient anymore: ops needs a single pane of glass view – a holistic view of all the services and their interconnections.

Now, let's go back to our example where we have an order that needs to be paid. The order status will change from not paid to paid and trigger the delivery. The status will only change if the bank has approved the payment, so the trigger to change the status comes from a service that's outside the service that actually holds the order. If the API between the bank and the order queue fails, the operators will need to know about this. So, monitoring should also include monitoring the API to the bank and checking if the bank service is *online*. Here, event monitoring across the full chain becomes key, as shown in the following diagram:

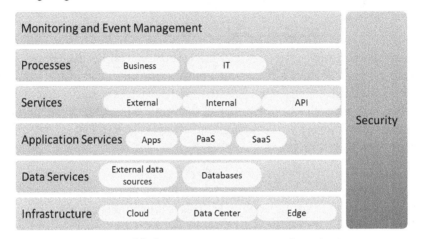

Figure 6.6 – Full chain event management and monitoring

This will certainly lead to more complexity in ops, and that's why we need to look for ways to automate this as much as possible. Microservices, SOA, and the event-driven architecture create more reliable systems: independent services simply wait for the trigger before the action is executed and the next service is started. Independent services allow for a fire and forget model: a system fires a trigger and then forgets about it; then, the system moves on to the next event and the next trigger. But we want to ensure that the trigger is actually received and maybe even check if the trigger is processed correctly.

In the next few chapters, we will learn how to monitor these processes and how we can automate event management in event-driven systems using microservices. This is exactly what AIOps does: making operations easier by means of AI and ML.

Planning operations with a maturity model

In this section, we will look at a maturity model for IT operations. Then, we will learn how to apply this to the enterprise and get it to continuous operations. Finally, we'll learn how to get it ready so that it can be implemented by AIOps.

The basic operations maturity model looks as follows:

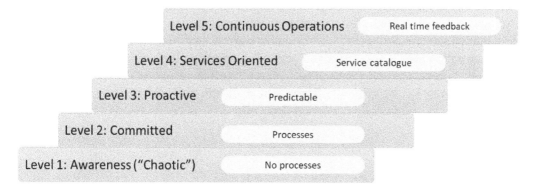

Figure 6.7 – Operations maturity model

The first level is sometimes referred to as chaotic. Processes are not documented here; operations are merely just firefighting. At this level, it's the tools that define how operations work, instead of having an architecture in place that also defines the toolset. Most enterprises have passed this level.

However, a lot of enterprises are stuck at the second level. This is the committed level, where processes are defined. Incident, problem change, and project management is in place, but the processes are only integrated in a very limited way. There's no overarching event management or *single pane glass view*. In other words, events will still trigger unexpected system behavior and might cause severe disruptions to the business.

Most enterprises will be somewhere between levels 3 and 4, depending on where they are in their digital transformation. The third, proactive level is already quite challenging to achieve. At this level, enterprises can analyze trends, have end-to-end event monitoring in place, have larger parts of their IT automated, and, most importantly, can predict events and taking proactive measures. This is the level where AIOps can play a significant role. We will learn more about this in *Chapter 7, Understanding the Impact of AI on DevOps*.

At level 4, IT delivery is fully defined as services to the business. This is the level where the event-driven architecture becomes relevant. Business functions are mapped to IT services. In fact, IT services can support business decisions. There's a well-defined service catalogue and all the processes are integrated, including cost management. The final stage in the maturity model in IT is real-time responsiveness to business events and innovations that are part of operations, thus adding value to the business. This is continuous operations stage, where we loop feedback to the development processes but in real time and in a fully automated fashion. Most enterprises will get there for some applications and some business functions, but hardly ever for the entire enterprise and all business processes.

The operations maturity model aligns with the **Capability Maturity Model (CMM)**, which also has five levels. Level 1 is the initial level, where processes are poorly controlled and unpredictable. At level 3, which is where most companies are at, the processes are clearly defined and well understood, allowing for proactive event management. Level 5 is the optimization level, where processes and delivery are continuously improved. As we mentioned previously, most enterprises will reach this level for certain processes and products, but rarely for the whole enterprise. The CMM model looks as follows:

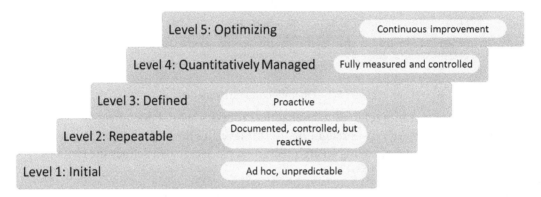

Figure 6.8 – The Capability Maturity Model (CMM)

Level 3 is the focus level: becoming proactive. In the next few chapters, we will learn how AI and ML can help us with that.

Summary

In this chapter, we discussed operations management and how this is changing due to digital transformations and the implementation of DevOps. First, we looked at the roles and responsibilities of operations and the different operational service management processes. We also discussed some trends that will impact operations in the near future. The overall conclusion we attained is that the role of ops will change, mainly because of digital transformation and the implementation of agile and DevOps. To become agile, we need operators to be able to focus on their distinctive tasks. We then discussed a demarcation model with product ops and platform ops.

Next, we learned how the architecture will change to become a more event-driven architecture and what the position of ops will be here. Ops will need to have a single pane of glass view, overseeing and even predicting events in the full chain so that proactive measures can be taken. This is what level 3 describes in the operations maturity model: proactive and predictable. AI and ML will help in this.

In the next chapter, we will discuss the impact of AI in the enterprise and IT operations.

Questions

1. Name three components that must be part of the technology architecture.

2. Name the four value streams that IT4IT defines for IT delivery.

3. An important component of the event-driven architecture is the principle of contained, independent operating services that communicate with each other using APIs. What do we call these services?

4. On what level in the maturity model would AIOps fit?

Further reading

-   *The New IT Operating Model for Digital*, by Gartner. Published on `https://www.gartner.com/en/documents/3770165/the-new-it-operating-model-for-digital`.

-   *Designing and Implementing the I&T Operating Model: Components and Interdependencies*, by Gartner. Published on `https://www.gartner.com/en/documents/3956725/designing-and-implementing-the-i-t-operating-model-compo`.

-   Blog on IT4IT by Rob Akershoek: `https://www.4me.com/blog/it4it/`.

-   *Solutions Architect's Handbook*, by Saurabh Shrivastava and Neelanjali Srivastav, PacktPub, 2020.

7
Understanding the Impact of AI on DevOps

In this chapter, we will introduce **artificial intelligence** (**AI**) and what the impact of AI is on DevOps. We will discuss how this is driving a shift left in operations, by enabling the fast identification of issues already at the beginning of the DevOps cycle, using AI and **machine learning** (**ML**). Before we can implement systems such as AIOps, we need to get the enterprise ready for AIOps in the first place by creating visibility of all IT assets and workflows and mapping them to AI-driven processes. Next, we need an integrated toolset for both development and operations. Leading public cloud providers offer native toolsets, as we will see in this chapter.

After completing this chapter, you will have a good understanding of the concept of AI in DevOps processes. You will also have learned how AI-driven systems can help in achieving shift left. Before we discuss the possible outcomes and benefits of AIOps, we need to create full visibility of all assets and processes in the enterprise's IT. In this chapter, we will also learn why that is important and how we can achieve full-stack visibility.

In this chapter, we're going to cover the following main topics:

- Introducing AI and ML
- Understanding the shift-left movement in DevOps
- Defining the first step – DevOps as a service
- Creating the IT asset visibility map
- Measuring the business outcomes of AIOps

Introducing AI and ML

In this section, we will briefly introduce the concepts of AI and ML. There have been complete bookstores worth of books written about AI and ML, but in this section, we will merely give a definition and describe how these concepts will change development and operations:

- **AI**: The broadest definition of AI is a computer technology that simulates human behavior. In most cases, AI is used to express the fact that software is able to react to events in an autonomous, intelligent way by deducting and analyzing and, by doing that, reaching decisions without human interference.

- **ML**: After AI is machines that learn how to perform tasks and execute actions by analyzing earlier events, and then use this experience to improve autonomous decision making. To enable this, both AI and ML as technology need data and they need to understand how to interpret this data.

AI and ML are not magic. You will need to define the scope for these technologies, just as with any other concept. Next, you will need to prepare environments to be ready for AI and ML. For example, an enterprise will need to have a good understanding of automation to start with and a complete overview of all of their assets. Otherwise, even AI will be working *in the blind*, bringing no value.

Introducing and implementing AIOps starts with a different mindset: improvements start with the early detection of possible failures and learning how to prevent them before they enter production, instead of detecting and correcting failures in production. This is the domain of shift-left thinking. We will learn more about that in the next section.

Understanding the shift-left movement in DevOps

Shift left has become a popular term over the past years. But what do we mean by this? It's about moving activities that were originally planned at a later stage up to the beginning of a process. This is typically the case with testing, which for a long time was executed as soon as the whole product was delivered to a test team. Shift-left testing has become an important paradigm in DevOps: executing tests as early as possible. By having tests already from the beginning of development, issues will be found much sooner and can be fixed in that early stage. It will improve the end product. The following figure shows the impact of shift-left testing:

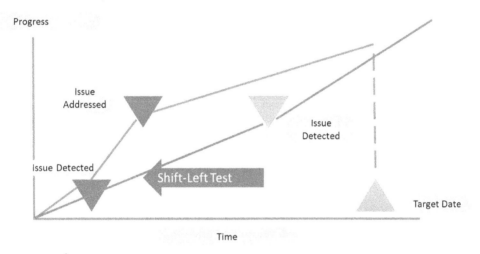

Figure 7.1 – Impact of shift-left testing

The shift-left principle can be applied to more processes in DevOps. Think of the very first step in DevOps: design. IT teams, both software developers and cloud engineers working on the infrastructure, should have a good understanding of the business requirements before they start building a solution. One of the major pitfalls in IT is that IT is building something without completely understanding these business requirements. A strong collaboration between business and IT can solve this by adopting other design approaches. Design thinking fits in perfectly and is a good example of shift left.

Design thinking starts with evaluating the perspectives of all parties involved in the development: in the methodology, this is referred to as *empathy*. The next step is to define the problem, brainstorming and generating ideas to solve the problem from every angle, then building and testing the prototype. Testing, however, is not the final stage. On the contrary: design thinking is an iterative process, just as DevOps. Products will get better with every cycle. The key is to involve IT already at the very beginning of the project, in the phase where business requirements are defined. The process is shown in the following figure:

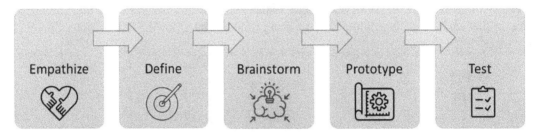

Figure 7.2 – Process of design thinking

Finally, shift left is also applicable to deployment and operations in DevOps. This is where automation, templating, and blueprinting play a major role. With automated templates, pre-approved patterns, and processes, we can shift deployment to an early stage. Using automation, we can achieve consistent deployment applications that will help operations in managing these environments.

Pre-approved patterns also include **test-driven development** (**TDD**), shifting testing all the way to the beginning of the development and deployment process. In *Chapter 3, Architecting for DevOps Quality*, we discussed TDD, where the team writes the test cases first and then the code. The code is written to the specifications of the test case, proving that requirements have been fulfilled.

In short, the shift-left principle is about reducing failures in an early stage, making end products more stable and resilient. Issues are often only discovered in production, typically caused by inconsistencies in the deployment of systems. Manual tasks or the use of a lot of different tools increase the risk of these inconsistencies. Developers using different tools than operations can result in issues that need to be fixed by manual tasks. In **Site Reliability Engineering** (**SRE**), this is referred to as toil, as we have seen in *Chapter 5, Architecting Next-Level DevOps with SRE*. Or, issues are caused by different procedures. Automation, templating, and TDD can avoid these issues occurring and reduce the failure rate. Templates, patterns, and blueprints are tested and improved continuously, leading to more stable operations.

AI and ML can help in all of this. First of all, AI-driven monitoring will help in detecting issues and especially inconsistencies at an early stage. It will learn from these inconsistencies and suggest and even implement improvements in code and procedures using ML. But before we dive into that, we need to discuss automation a bit more as part of the shift-left paradigm, shifting as much as possible to cloud services using integrated toolsets, implementing DevOps as a service. That's the topic for the next section.

Defining the first step – DevOps as a service

Consistency is the key to success. That applies to almost anything and it certainly applies to DevOps. Dev and ops need to collaborate in the same toolset: that is what DevOps as a service is about. DevOps as a service enables shifting left, but is also a good starting point for implementing overarching monitoring systems, including AIOps.

> **Note**
>
> AIOps is way more than just a monitoring tool, as we will find out in the following chapters. However, AIOps starts with the monitoring of complex environments. By gathering data from these systems and analyzing this, it will be able to track and remediate systems and processes, including the automation of repetitive tasks. AIOps is capable of discovering patterns for which it can define automated triggers. But it can't do this if it can't monitor the source systems.

DevOps as a service will track every step in the development and delivery process, but the real value is that it provides feedback as soon as an issue in that process is detected. The value lies in the fact that this feedback is already collected before the software is pushed to production. From the start of the development cycle, an integrated toolset enables the tracking of bugs and errors and sends this back to the development team, way before operations is confronted with faulty software and unpredicted behavior from systems. This is a true shift left: shifting things that we typically do at a later stage to the beginning.

DevOps as a service thus represents an integrated toolset that enables collaboration between developers and operations. The tools have to cover all steps in the DevOps process and basically work together as one tool. Cloud platforms provide these tools. In this section, we will discuss these tools in Azure, AWS, and **Google Cloud Platform** (**GCP**):

- **AWS**: AWS CodeBuild, AWS CodePipeline, and AWS CodeDeploy are the three main solutions to look at. CodeBuild is a managed service for building, compiling, and testing code through automated processes. CodeBuild also provides unique encryption keys for every artifact that is built and stored in the code repository. Deployment scenarios are defined in CodePipeline, up until production, where CodeDeploy enables the delivery to targeted infrastructure in production. CodeDeploy also takes care of patching, upgrades, and the synchronization of builds.

- **Microsoft Azure**: Azure DevOps is the integrated toolset in Azure for development and deployment. It's a sort of Swiss Army knife: it acts as one tool, but under the hood, it holds different solutions that work together. You can manage the code in Azure Repos, which provides support for Git repositories. Building, testing, and deploying code is done in Azure Pipelines. More extensive testing can be executed using Azure Test Plans. Next to this, Azure DevOps provides Azure Boards, which is used to track projects: it can be compared to Kanban boards. Finally, it provides Azure Artifacts, where developers can share NuGet, NPM, Python, and Maven packages from other sources into Azure DevOps.

- **Google Cloud**: Google Cloud offers the Operations Suite, formerly known as Stackdriver. The most interesting part for developers is likely Cloud Debugger, which allows analyzing code in a running state and finding bugs without stopping the applications. Code deployment is done through Deployment Manager. GCP also offers a powerful tool for fast and automatic issue detection and analysis with Cloud Trace—in fact, this is already very close to AIOps.

Having integrated toolsets will help us in the shift-left movement and create a good starting point to implement AIOps. But we need to do one thing first and that's making sure that we have visibility of every asset in our IT environment. That's the topic of the next section.

Creating the IT asset visibility map

There's a famous line in *Alice in Wonderland*: "If you don't know where you are going, any road will get you there." You can actually turn this around: if you want to go somewhere, you need to know where you're coming from. Let's put this into practice: if we want to transform the enterprise, we need to know what it is we are transforming. That's why every approach to digital transformation starts with assessments and discovery. An enterprise needs to have full visibility of all of its assets. The following figure shows the basic steps in a migration and transformation plan, starting with the assessment:

Figure 7.3 – High-level plan for migration

When all assets have been identified, we can start the planning of migrating and transforming these assets to a new target landing zone, typically a platform in the public cloud. Applications need to be validated so that the right strategy can be defined: rehost, replatform, rebuild. This is the domain of app modernization that we discussed in *Chapter 4, Scaling DevOps*. The final step is to plan the migrations and transformations in waves. A *big bang* can be a strategy, but in large enterprises, this is certainly not recommended.

Getting back to the reasons why enterprises adopt Agile and DevOps: enterprises do this to speed up the delivery of products and become more flexible so they can respond faster to the changing demands of customers. To gain that speed, they need to rely on stable systems and operations, so time can be spent on development instead of fixing issues. IT needs to become more predictive and actually avoid issues occurring. Data coming from assets is crucial. That data needs to be collected and analyzed in real time.

The first source of this data is the **configuration management database** (**CMDB**). The problem with a lot of CMDBs is that the information they hold is not accurate. The root cause for that is the fact that a lot of data is still entered manually by, for instance, importing spreadsheets; monitoring and asset collection is not done in real time; or data is scattered across multiple CMDBs that need to be synced.

Next, CMDBs are not very often *cleaned*, so they contain noise. This typically is a result of not updating the CMDB after changes have been executed. The CMDB should be the single source of truth when it comes to capturing assets, but without real-time updated—automated—information, this becomes a challenge. The CMDB will not reflect the actual status of systems anymore.

You will have noticed that we're using two different terms in this section: assets and configuration. These are different things, and they are equally important in understanding how the enterprise's IT is set up. Only when we have full visibility of assets and configurations can we start planning migrations and getting the tools in place that will help operations to become more predictable. Ops needs a system that does the following:

1. Knows what systems are in the IT environment
2. Knows how these systems relate to each other
3. Tracks the status and configurations of these systems in real time

We start with knowing what systems we have in our environment: the asset visibility map. This is asset management: basically a list of every physical and virtual system, including the software that is used and licenses so that we know when systems will need to be upgraded or replaced, for instance, because software gets to the end of support or licenses have expired. This is addressed by the life cycle management process.

We also need to know how these systems are configured and what their relation is to other systems, including dependencies so that operators know what the impact is when a database is shut down. This is configuration management.

So, full visibility involves not only all assets but also a comprehensible visualization of connections, dependencies, and processes. Without this information, operations will have to spend hours finding the root cause of a problem and learning how to solve it.

Working in the cloud makes real-time asset collection easier. As an example, with the `Get` command in Azure, you can create lists of assets in Azure subscriptions. Again, real time is crucial here. Since cloud systems tend to change fast, it becomes even more important to have instant, accurate asset data. How do we create this visibility? We define five layers, as shown in the following figure:

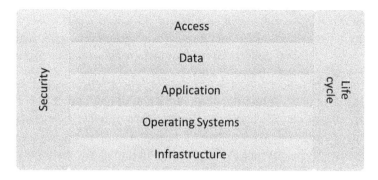

Figure 7.4 – Layers of asset management

Let's explore these five layers in a bit more detail:

- **Infrastructure**: This includes virtual machines and all network components.

- **Operating systems**: Are all operating systems up to date? But also: how are they configured? For example, enterprises typically have security standards, applied to images of operating systems that are hardened with these standards. Are all systems implemented with that same image? Systems that have been set up with another image might be vulnerable.

- **Application**: What application software has been installed and what version? Is software properly patched and licensed?

- **Data**: Where is data stored and how is it stored? For example, is data encrypted and in what way?

- **Access**: Who or what has access to the four other layers?

> **Note**
>
> There are two vertical layers in the figure. Life cycle management is applicable to the entire stack. Are all components still compliant, properly licensed, and not running out of service? This is even valid for the access layer: think of accounts that are not used anymore and should be disabled.
>
> Security is intrinsic to all layers. It's not something that we only have to take care of in the infrastructure or the data layer. Every layer needs to be compliant with the security policies. We will learn more about this in the third part of this book when we will talk about security in DevOps.

But enterprises will likely have more than only assets in a public cloud. Also, assets in the public cloud might have relations and dependencies with even non-cloud systems. So, we need an overarching system that collects all data and where this data is maintained: the single-pane-of-glass view or full-stack visibility.

Azure, AWS, and GCP offer APIs to **enterprise service management** (**ESM**) systems such as ServiceNow and BMC. ESM goes way further than only IT: it correlates business processes to IT and provides one view of how IT supports these business processes. These systems allow predicting how changes in IT systems will impact the business processes.

So far, we have only talked about the assets and configurations that reside within the enterprise. But in this fast-changing world, the enterprise will also have a massive amount of external sources that create data. Think of data coming from websites and social media platforms. These might be real assets to a company, but data from these sources are likely to have an impact on business processes. Therefore, enterprises will also have a need to analyze this data.

For example, a campaign on social media might raise sales, requiring the extra capacity of the sales systems, including the company's websites. If this is not foreseen and systems crash because of traffic overload, it will definitely cause damage to the company that is not limited to loss of revenue, but also reputational damage.

Predicting the behavior of systems and measuring the impact on a business requires full visibility of the entire ecosystem of the enterprise's IT. AIOps will help and in the last section of this chapter, we will discuss how.

Measuring the business outcomes of AIOps

In the previous sections, we discussed shift left and saw how we can define DevOps as a service. Next, we learned how to create total visibility of all assets in the enterprise as a starting point to implement AI-driven processes that will help improve development, deployment, and operations and with that, accelerate a shift left in IT. How will AI help with that?

- AI is about analyzing data. AIOps is no different: it analyzes operational data and is able to give recommendations on improving systems in terms of performance and efficiency.

- To get valid data and recommendations, AIOps has to reduce noise. Noise is a very common problem in operations and specifically in monitoring systems and CMDBs, as we learned in the previous section. What is really an issue and what is a false alert? AIOps is capable of analyzing these alerts and, with the help of algorithms that group alerts, can identify and prioritize these. The outcome is that this saves a lot of time in operations: operators can now focus on alerts that really impact systems—and therefore business.

- A very strong capability of AIOps is that it can correlate alerts and events to identify the root cause of an issue. A prerequisite is the full-stack visibility that we discussed in the previous section. AIOps are learning systems: meaning that they will first need to capture all the assets and understand the relationships between them.

- One more time: AIOps are learning systems, so they will learn what the normal behavior of systems is. They will then recognize anomalies and correlate these proactively with the possible business impact. Take the example we used with the campaign leading to a spike in sales. AIOps systems will detect abnormally high traffic much faster than operations and trigger scaling more quickly.

- Triggering scaling implies that AIOps is highly automated. AIOps can be used to take care of routine tasks, such as executing backups.

How can we measure the benefits of adding AI? The following **key performance indicators** (**KPIs**) can help:

- **Mean time to detect (MTTD)**: How much time elapses before an issue is detected? AIOps will learn how to detect failures, analyzing patterns and the behavior of systems. AIOps will also know how *severe* an issue is by analyzing how it affects business processes. Since AIOps uses ML, it will learn how to detect issues faster, how to predict them, and eventually avoid them by proactive recommendations.

- **Mean time to acknowledge (MTTA)**: This logically follows MTTD. MTTD is about detection, while MTTA is about how fast AIOps can route the issue to the right operators. This is covered by automation of the workflow process: AIOps will first recognize the issue, determine the impact, and then decide which operator it should be routed to for further investigation. This includes raising the incident to the highest level of criticality when, for instance, crucial business processes are impacted. AIOps may decide to flag the issue as critical, triggering the crisis workflow. Using these systems, a lot of time will be saved in comparison to manual intervention.

- **Mean time to resolve (MTTR)**: Using AIOps, it can be quickly identified whether similar issues have occurred before and what solution was executed to mitigate them. In short, AIOps will help in finding and analyzing solutions fast, restoring the service as soon as possible.

- **Detection by monitoring**: A useful KPI to measure the success rate of AIOps is how many issues have been detected by AIOps before users actually noticed a drop in performance or even the outage of systems.

- **Remediation by automation**: This is sometimes referred to as *automate automation*. AIOps will learn what solutions are used to solve issues. Sophisticated systems will also learn how to automate these solutions, proactively taking measures to prevent the issue from happening again. A useful KPI to measure the effectiveness of AIOps is to track how many remediating actions are automated by AIOps and what the effect is on the availability of systems.

Be aware that AIOps is still in its very early stages. Quite a number of tools that call themselves AIOps aren't able yet to, for instance, "automate automation," as described in the last bullet. However, AI and ML will absolutely evolve and get more mature in the coming years. It will be a first step in implementing AI and ML to DevOps and IT as a whole and changing the software development by doing the following:

- Creating prototypes

- Automated detection and analysis

- Automated correction

- Automated code generation

- Automated testing

This chapter introduced the concepts of AI and ML and how these technologies will impact the development, deployment, and management of IT systems. AI will help in creating more reliable systems and improve the development of software. AI will certainly not replace developers or operators, but their roles might change as they learn to work with AI-driven systems such as AIOps. First, we will discuss how we can integrate AIOps into our architecture, which is the main topic of *Chapter 8, Architecting AIOps*.

Summary

After a short introduction to AI and ML, this chapter discussed how these technologies will help in making better software and more reliable systems. AI enables the shift-left movement: shifting things that were typically done in a later stage to the beginning of the development and deployment cycle. With AI, it's possible to detect issues in a very early stage and by means of automation, AI will also be able to trigger correcting actions.

Since AI and ML are learning systems, they will learn how to predict and possibly prevent issues from happening. For this, AI needs real-time data coming from source systems, hence the first step is to get a total overview of all assets in our IT environments and make sure that these systems are monitored, providing real-time logs. We learned how to create this full visibility using five layers.

In the last section, we discussed KPIs used to measure the outcomes of AI-driven systems. Although AIOps is still relatively new, the technology is very promising in getting better insights into the behavior of IT and predicting the impact of IT events on the business. In the next chapter, we will learn how to integrate AIOps into the enterprise architecture.

Questions

1. Design thinking is a method to create a shift-left movement. Design thinking starts with evaluating the perspectives of all parties involved in the development. What is the term that is used to describe this step in the methodology?

2. AWS offers DevOps as a service using native tools. What are the three tools for building the code, planning the deployment scenarios, and the actual deployment to production instances?

3. What does MTTA stand for?

Further reading

- *AI Crash Course*, by Hadelin de Ponteves, Packt Publishing, 2019

- Blog by Clive Longbottom: `https://searchitoperations.techtarget.com/definition/DevOps-as-a-service-DaaS`

- *Azure DevOps Explained*, by Sjoukje Zaal, Stefano Demiliani, and Amit Malik, Packt Publishing, 2020

8
Architecting AIOps

In this chapter, we will learn how **artificial intelligence for IT operations** (**AIOps**) platforms are designed and what sets them apart from any other monitoring tool. You will get a better understanding of why these platforms will become a necessity in the future of IT. The chapter starts by explaining the logical architecture model and then drills down to the key components in AIOps using data lakes and analysis through machine learning. We will define the reference architecture for an AIOps platform and learn how this will drive operations as we integrate the technical services architecture in AIOps. The chapter will provide insights into algorithms, anomaly detection, and auto-remediation.

After completing this chapter, you will be able to recognize the various key components of AIOps and define the reference architecture. You will learn some important lessons in implementing AIOps in enterprises, avoiding some major pitfalls.

In this chapter, we're going to cover the following main topics:

- Understanding the logical architecture
- Defining the key components of AIOps architecture
- Integrating AIOps with service architecture
- Defining the reference architecture for AIOps

Understanding the logical architecture

Before we dig into the architecture of AIOps, we need to understand the structure of logical architecture. In the previous chapter (*Chapter 7, Understanding the Impact of AI to DevOps*), we learned that one of the first steps to get started with the implementation of AIOps is to get full visibility of all our IT assets and processes. This is a requirement to *feed* the AIOps model with the basic information of how the IT environment looks. Most AIOps tooling will scan the environment using agents, but that is not sufficient. It also needs to understand the relationship between assets and what process flows are implemented. That is covered in the logical architecture.

The logical architecture is not about technology. It doesn't care what type of machines are used or what software. Logical architecture describes systems without the definition of the underlying technology. It describes the relationship between logical components in a system. The choice of a particular technology—infrastructure, software, code—is added later to this architecture. A logical architecture is meant to provide a comprehensible overview of the system as input for the technical design. Again, a logical architecture typically consists of layers, as shown in the following diagram:

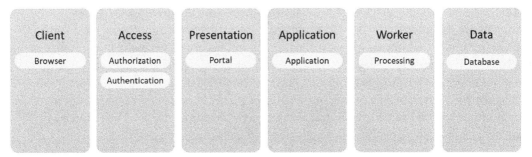

Figure 8.1 – Components of a logical architecture

We will discuss these layers and the components in more detail later in this section.

The logical architecture should be the starting point for any architecture. It holds every component that a system needs to fulfill its function. So, from the desired functionality, the architect must work out the different components that a system needs to be able to meet this functionality, without being bothered by constraints that inevitably will come with the choice of technology. The architect will need to think about the various system layers, the components in these layers, and the interaction between the layers. Next, they will also need to consider the interactions with other systems.

Only when this logical model is completely worked out in every detail, the physical architecture can be defined, translating the logical components into technical components such as infrastructure with network devices and servers, operating systems, and finally, the technical application components, including the interfaces and APIs.

To help the architect in defining the logical architecture, it's recommended to work in tiers as shown in the previous figure. Tiers include client, access, presentation, application, worker (in some literature referred to as business), and data. It follows a common pattern that systems use to perform a transaction.

> **Note**
>
> A transaction doesn't have to be a financial transaction. In system architecture, a transaction refers to an action that a system performs, triggered by input and leading to a certain output, or a request leading to a response. Of course, this can be a real transaction, such as an ordering process. But a transaction could also be a message that is transferred to another system.

How would the request-response cycle look when we break it down into separate steps?

1. The user raises a request through the presentation tier.

2. The user is authenticated through the access tier.

3. After authentication, the request is transferred to the application tier.

4. From the application tier, the request is processed through the worker tier.

5. The required data is collected from the data tier.

6. Data is processed through the worker tier.

7. A response to the request is prepared in the worker tier and sent to the application tier.

8. The response is transferred from the application tier to the presentation tier.

The request-response flow is shown in the following figure:

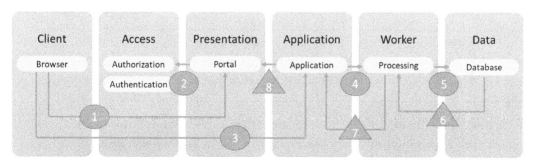

Figure 8.2 – The request-response flow in a logical architecture

The circles represent the request and the triangles show the response flow. Now, let's look at the different components in more detail:

- **Client**: This is the entry point for the user to start using an application. Typically, this is a browser or an applet on a device that allows the user to connect to applications.

- **Access**: We don't want just anyone using systems, so users need to be authorized and authenticated. The access tier sits between the client and the presentation tier: users use the client to get to the presentation tier and are then authorized and authenticated in the access tier. After access is granted, the request is processed.

- **Presentation**: The presentation tier has two main functions. First, it helps the user to put in the request in a comprehensible way. Once the request has been processed, the response is presented in this tier.

- **Application**: This is the tier where the application technology is stored.

- **Worker**: Sometimes this tier is referred to as the business tier, but that might be a bit confusing. This is the tier where the request is processed. Basically, this is the tier where the application functionality is stored. It holds all the components that enable an application to prepare a response to the user's request, including the APIs that, for instance, collect the required data and process that data.

- **Data**: This tier should only be used to store the data. The worker tier sits between this data tier and the presentation tier so that clients can never access data directly. They will always need to go through the worker tier.

This is straightforward if it concerns one system, but enterprises have more systems. These systems connect to each other and might very well depend on other systems. If an enterprise has implemented **single sign-on** (**SSO**), a user might only have to go through an access tier once. But it's very likely that the request will follow a workflow through different worker tiers belonging to different systems. There might be data required that comes from various data sources. It's certainly not a given that the user gets the responses presented in just one presentation tier, although that might be very much desired from a user's perspective. If we think about that, the logical architecture becomes already quite complicated, let alone if the architect tries to get in all the technical details at this stage.

AIOps works with this logical architecture. From this, it understands how systems are designed and what relationship they have to other systems. This is the basis for the reference architecture of AIOps. Before we get to that reference architecture, we will look at the different components of the AIOps architecture. This is the topic of the next section.

Defining the key components of AIOps architecture

In this section, we will discuss the key components of AIOps. Then, we will look at the operating services that provide input for AIOps, and finally, in the last section, we will draft the reference architecture.

First, let's recap why we should have AIOps in IT and specifically DevOps. IT has become more complicated. Systems are hosted on various platforms, connecting to other systems, using a lot of different data sources, with an equal amount of variety in data formats. We have come to the point where it has almost become impossible for a human to maintain an overview of these complex landscapes. Yet, the market is demanding new features at a rapidly increasing pace. Developers are under a lot of pressure to deliver code that fulfills new functionality, while operators need to make sure that systems are running stably. AIOps can help you do the following:

- **Process and evaluate data**: AIOps can rapidly, in almost real time, process data coming from different systems and validate this data. By doing this, AIOps can generate alerts and suggest mitigating actions to maintain the desired performance of systems.

- **Rapid root cause detection and analysis**: Because AIOps has access to all that data and it understands how systems are interconnected, it will be able to get to the root cause of an issue much faster than operators.

- **Enable automation**: By processing and evaluating data, AIOps will recognize patterns. These patterns are translated into automation sequences, eliminating manual tasks. Also, AIOps will learn from these patterns by analyzing the outcomes of certain processes. If a pattern constantly leads to an undesired outcome, it will come up with suggested actions to change the pattern. Engineers will benefit from these suggestions. They will have to spend less time searching for issues in the systems and more time improving them.

- **Anomaly detection**: Because AIOps learns patterns, it will also be able to recognize deviations in these patterns. These deviations might be anomalies: events that need attention. That could be a user trying to get access within minutes from two locations that are separated miles from one another to the unexpected trigger of an API from the CI/CD pipeline. AI therefore will also greatly benefit the security of systems, something that we will discuss further in the third part of this book, about DevSecOps.

- **Automated remediation**: Sophisticated AIOps systems might be able to autonomously solve issues, since they have learned patterns and are able to detect anomalies. Since AIOps systems are highly automated, they can also learn how to execute remediating actions in the event of *known* issues, typically issues that have occurred in the past. AIOps will learn how these issues have been solved and in that case, the action can be automated. Be aware that the conditions of these issues need to be similar to the conditions when the issue first occurred.

Overall, AIOps will definitively increase the efficiency of IT and more particularly the DevOps cycle. Developers have started using AI to discover correlations and patterns in various systems, helping them to improve the code, including predictive analysis. It means asking the *what if* question repeatedly, only much faster and with a far better chance of getting more accurate answers.

To be able to do this, AIOps systems require several components:

- **Source repository and control**: As with CI/CD in DevOps, all code has a source so that systems can always be restored from that source. All designs, patterns, configuration parameters, and application code, and all other IT artifacts, are kept in a source repository that is brought under version control. This is the *baseline* for all systems. The **Configuration Management Database** (**CMDB**) that we discussed in *Chapter 7, Understanding the Impact of AI on DevOps*, is part of this repository but will likely not hold application code. So, the full repository will contain various sources.

- **Data Repository for Big Data**: AIOps systems collect a lot of data from various sources that they analyze and use to train themselves, modeling the data into patterns to enable further automation. This data is collected in its original formats in distributed file structures, such as a data lake.

- **Monitoring**: AIOps systems might use agents to track how systems perform and if connections between system and system components are available. Most systems, however, will aggregate data from other monitoring sources. In a public cloud, this could be Azure Monitor, AWS CloudWatch, or Google Cloud Monitoring, combined with other, specialized tools or vendor-specific tools to monitor specific components in the IT environment. AIOps will get the raw data from these tools and use this to train itself.

- **Automation**: Automating repetitive tasks and for auto-remediation.

- **Machine Learning**: This is the core of AIOps. From the source repository and the collected data, it will learn how systems behave and connect to other systems. AIOps will learn how applications and data in applications are used and how requests are processed. AIOps systems will use the big data, compare it to source repositories, and continuously learn from these analyses in order to automatically trace issues and come up with suggestions to mitigate or improve, providing feedback to developers and operators.

Now we need to understand how these components work together. IT research and advisory company Gartner has defined a model for AIOps that consists of three domains:

- **Observe**: This involves real-time tracking and tracing of all metrics across the systems. AIOps monitors performance, connections, and anomalies by correlating and analyzing data with historical data coming from source repositories.

- **Engage**: This involves all data that is related to incidents and changes. The engage domain integrates **IT service management** (**ITSM**) processes with AIOps. This will be discussed in more detail in the next section.

- **Act**: This involves the automation and any action that AIOps systems execute based on the monitoring data (observe) and the output of processes (engage).

The model is presented as follows:

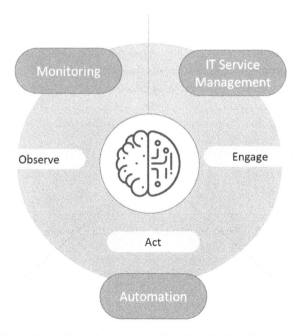

Figure 8.3 – Three domains in AIOps (original by Gartner)

To be able to act, AIOps needs to understand and learn how IT services are delivered through the enterprise. We will discuss this in the next section, where we learn how technical service architecture is integrated with AIOps, driving operations in a more efficient way.

Integrating AIOps with service architecture

So far, we have looked at the logical and different components of AIOps. One of the main reasons to bring AI into operations is to relieve operators of a lot of manual tasks by proactively monitoring systems and even mitigating possible risks before they actually materialize. In other words, AIOps is invented to reduce what **Site Reliability Engineering (SRE)** calls toil. That's the whole point of implementing DevOps and AIOps: to reduce toil and create time to continuously design and build better systems.

But it's not only about the systems themselves. Enterprises also have processes to deliver services. That's the domain of the service architecture. Or, better and more precise: the technical service architecture. That includes the processes that are discussed in the following subsections.

Monitoring

This is not something that simply comes out of the box. On the contrary, architects need to define a monitoring architecture including what business functions, applications, and underlying infrastructure must be monitored in order to get accurate and appropriate data about systems' conditions at the earliest stage possible. The next question is how monitoring can be automated: what thresholds must lead to actions, and can these actions be automated as well? What logs must be stored, where, and for how long? It's recommended to store logs automatically, as part of the big data that AIOps can use to train itself. Monitoring architecture next defines the routing of alerts: this is crucial input for AIOps platforms.

Problem management

This starts with an incident, preferably detected through monitoring rather than a user reporting an issue to a service desk. Not every single incident will lead to a problem, though. A user who can't log in to a system because their credentials have expired might become an incident but rarely will lead to a root cause analysis and/or a problem. If incidents are severe, impacting parts of or even the whole business, then problem management needs to be executed. The same applies when similar incidents occur multiple times. Problem management is needed to analyze the cause and to determine whether there are correlating events and patterns. Problem management therefore requires a deep knowledge of systems and tools. With today's complexity of the enterprise IT landscape, the execution of problem management will inevitably lead to analyzing a vast amount of data coming from various systems. AIOps will help in analyzing this data, determine trends, and get to root causes faster.

Configuration management

What is the desired state of a system? That is the main question in configuration management. **Configuration items** (**CIs**) are kept in a repository, including the status of their life cycle. CIs include every component that forms a system: network connections, servers, software, APIs, and policies. Each of these components must be configured as defined in the architecture to maintain consistency.

Example: the architect decides that a Windows server always runs the version Windows Server 2019, but with specific settings that are captured in a server policy. They also decide that a database server always runs SQL or uses Azure SQL as a PaaS service. As part of the architecture, a database server is never allowed to communicate directly with the outside world, so the architecture will also have policies on how database servers will communicate with other servers, including firewall rules and connection strings.

Configuration management needs to make sure that these configurations are kept during the entire life cycle of components, also when these components are migrated to different platforms or different stages in the promotion path (development, testing, acceptance, and production). For that reason, accurate configuration management is critical for ensuring system quality and good change management. Configuration management and desired state are the baseline input for AIOps.

Change management

Whenever systems or components of systems are moved, updated, upgraded, altered, or decommissioned, change management is involved. In other words, change management does not only take care of the implementation of new systems. Any time a system is changed in whatever way, change management plays an important role. To enable change management, it needs a single source of truth in terms of CIs and the configuration of components – the desired state. It needs to rely on configuration management, accurate monitoring, and consistent reporting of events that have been logged from systems. In AIOps this is identified as historical data, to which it compares the changes and validates whether systems are still performing well in the desired state. Anomaly detection will help in identifying issues as changes are implemented.

All these processes discussed so far result in data, stored in tickets that are raised by users or preferably events logged by systems. In the model that we discussed in the previous section about defining the key components for AIOps architecture, Gartner refers to this type of data as engagement data.

Engagement data is used for the following:

- Task automation
- Change risk analysis
- Performance analysis
- Knowledge management

Engagement means that ITSM communicates with operational data that is derived from monitoring, events, and system metrics. All this data is aggregated in one repository, a big data platform so that the AIOps system has one data library with which it can analyze system health and performance.

Big data in AIOps

This big data platform is the heart of AIOps: it will capture data from different systems without the silos these systems are in. What do we mean by that? A system can have components in a privately owned data center and communicate with PaaS or SaaS services in a public cloud. To get an integrated view of the entire system, AIOps needs to have access to one, aggregated data source – the big data platform. That platform will hold historical data, data on iterations of the system, and real-time system data using interactive big data analytics.

Recapping, AIOps platforms have operational data and process data:

- Data originating from many different systems

- Capabilities to analyze streaming, real-time data, and historical data

- Process data from events, logs, alerts, and root cause analysis

- Capabilities to integrate process IT workflows and correlating them with operational system data

- Capabilities to visualize data in a comprehensible way

The following figure shows the AIOps stack:

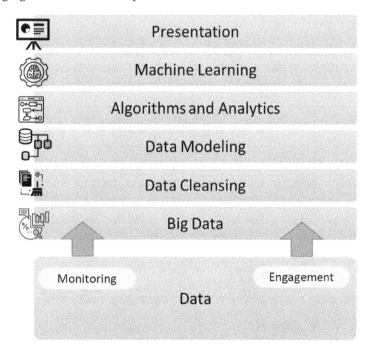

Figure 8.4 – High-level modeling of the AIOps stack

We have extensively covered the domains observe and engage in the Gartner model. Based on the data and the services patterns AIOps can now start executing activities, the final stage in AIOps. It can automate tasks by using runbooks and scripts. The final step is to translate the components in an architecture for an AIOps platform. In the last section, we will define a reference architecture, putting it all together. Lastly, we will provide some tools that will get you started with AIOps.

Defining the reference architecture for AIOps

In the previous sections, we studied the logical architecture of systems, the components of AIOps, and the technical service architecture. All these building blocks are used to define the architecture for AIOps. In this section, we will look at the reference architecture for AIOps.

First, let's recap the goal of AIOps. In *Chapter 7, Understanding the Impact of AI to DevOps*, we discussed the **Key Performance Indicators** (**KPIs**) for AIOps:

- **Mean Time to Detect (MTTD)**
- **Mean Time to Acknowledge (MTTA)**
- **Mean Time to Resolve (MTTR)**

AIOps adds artificial intelligence to IT operations, using big data analytics and **machine learning** (**ML**). The AIOps system collects and aggregates data from various systems and tools, in order to detect issues and anomalies fast, comparing real-time data with historical data that reflect the original desired state of systems. Through ML it learns how to mitigate issues by automated actions. Eventually, it will learn how to predict events and avoid them from impacting systems.

To enable this, AIOps works with predictive alerts based on algorithms and anomaly detection. It *knows* how systems should react and respond and how systems interact with other systems. Deviations are quickly discovered, analyzed, and validated: are these deviations in the range of normal behavior or are they outside that range? What could be the consequence if a system is showing anomalies?

AIOps can only value these events if it's provided with data: system data and process data, such as requests and data on incidents and problems that have occurred. It needs that data to learn and adapt the algorithms, so it becomes more accurate over time.

In the previous section, we learned about the three domains in AIOps: observe, engage, and act. Now we need to translate this into a reference architecture. We can do so by creating four layers to perform operational tasks. This is shown in the following figure:

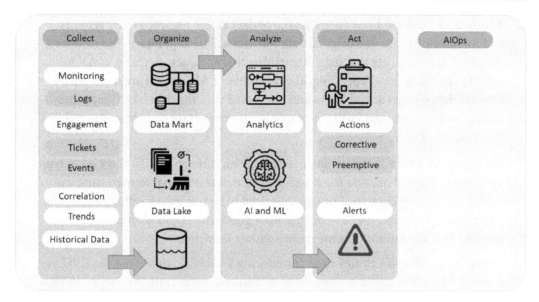

Figure 8.5 – Logical reference architecture for AIOps

These four layers contain various components, just as we saw in the logical architecture of systems that we discussed in the *Defining the key components for AIOps architecture* section of this chapter:

- Collect
- Organize
- Analyze
- Act

The process starts with the different services as input: requests, incidents, problems, and changes. In the previous section, we learned that this is included in engagement data. AIOps systems will now perform several steps:

1. Collect is the first step that an AIOps system will perform: it will collect observations and the engagement data from as many sources as available: system metrics, events, logs, and even information from tickets that are raised by a service desk. All this data is sent to a data lake, the main repository of the AIOps system.

2. The data lake will hold raw data: the source data as it is retrieved from systems and monitoring tools. The original data formats will be kept intact. The next step, therefore, is to organize that data. Remember that accuracy increases as AIOps has access to as much data as possible. The outcomes of AI analyses are more reliable as more data sources are added to and integrated with the data lake. That comes with a challenge: raw data will contain noise. It's important that this imported data is cleaned.

    An important lesson in data science is garbage in, garbage out. To allow AI to come up with accurate advice and *learn* with the right data, we need to make sure that it only uses proper datasets. Data models need to be trained as well. This will take some time: the system needs to be told what good data is and what it can be used for. This way, algorithms become more accurate over time.

    An example will make things clear. Monitoring will collect data from the CPU in a server. That CPU might have peaks of 90 percent, but overall, the usage will be around 20 percent. AIOps will need to learn that a value between, for instance, 20 and 40 percent is normal. It will also need to learn that a peak of 90 percent is OK if it's for a very short time and that it doesn't have to create an alarm for that. Now, the model can filter event and log data based on these parameters and identify systems that are running above that threshold. That's indeed something that regular monitoring systems already do, but AIOps will be able to correlate anomalies in CPU performance with other events. And it will be able to initiate remediating actions automatically.

3. Raw and cleaned, structured data is used to train the AI models, using machine and deep learning.

4. Real-time data is compared with historical data such as original configurations, executed changes, and passed events. By constantly comparing this data, the AIOps systems learn and will even be able to present a timeline of events. This allows operators in some cases to *time travel* through a series of events, investigating when and why a system state changed.

5. The final step is act. Trained AI models will send predictive alerts and start mitigating actions, based on the learning of previous events and actions that have been taken when a specific event occurred. Obviously, all alerts and actions are looped back as data into the data lake.

We are ready to start the implementation of an AIOps platform. In the next section, we will discuss the approach to a successful implementation and also identify some potential pitfalls. In the last section, we will give a short overview of some popular tools.

Successfully implementing AIOps

AIOps is still a rather new phenomenon in IT operations and implementing AIOps platforms can be very complicated. However, with the following steps in mind, it should be doable to perform a successful implementation. The first and probably the most important lesson is to start small, as with DevOps and SRE.

Before you start, you need a plan. Identify which systems would suit a pilot. Pick a system with a basic architecture: an application with just a website as the frontend, the application itself, and a database. Validate whether all assets belonging to this application are documented well and are identified in the CMDB. Next, ensure that you have a good – documented – understanding of workflows and processes that are related to this application.

Why is this important? The AIOps platform, but also engineers – operators – need to learn how to work with AI, ML, and analytics. What do operators want to get out of the platform? What type of alerts? And is it clear what they should do with the alerts or how they can use AIOps to analyze the alerts and help it in automating actions? Operators will need to learn what type of data the AIOps system needs, so they will need to learn about data requirements, data flows, storing data in data lakes, and modeling data analytics. This will definitely take time.

Now the AIOps system can be further configured, and more components of the IT landscape can be added, making sure that the system learns from the different use cases. It is highly recommended to run the platform in test mode for the first weeks or even up to 2 months, allowing AIOps to learn. Once operators are satisfied with the functionality and the outcomes of analysis, it can be pushed to run in production. But you should always validate outcomes with the original system metrics and the goals that were set to implement AIOps. The simple question is: have these goals been met in the testing phase? If the goal was to speed up the time needed to define the root cause of an incident, then that goal should be measurable and compared to the original metric. To put it differently: any implementation of a system must have a valid business case.

So, the first step is capturing operations data: logs and metrics from infrastructure components, applications, databases, and APIs. Next, we need data from workflows and also business processes. We might also need other relational data, for instance, data about sales and market sentiments, "predicting" when certain services are popular in specific periods and under what conditions.

Use case

An international operating railway company implemented AIOps a few years ago. Their infrastructure was built on various platforms, including AWS and Google Cloud, using a lot of different tools for monitoring and development. The lack of real-time visibility of the systems on these platforms and their dependencies slowed down developments, but also caused heavy delays in root cause analysis. This caused some major outages, impacting travelers. By bringing data together from the various platforms and integrating monitoring tools such as AWS CloudWatch and Google Analytics, the company was able to have an integrated view of all the correlated systems. AIOps tooling was implemented to analyze incidents in real time using this data, resulting in significantly lower MTTD and MTTR.

The use case shows that the more data we can get, the more accurate our AIOps platform will be. This data will tell how events correlate and eventually ensure that AIOps understands the root causes of these events. From this point onwards, you can train the platform on resolutions using real-time observations through monitoring and creating automated responses.

One crucial step in implementing AIOps platforms is to really involve operators. They are and will be very important to get real value out of the platform. In other words, AI is not about replacing operators – on the contrary. The system will enable them to get rid of tedious, repetitive, usually manual tasks – remember *toil* in SRE – and allow them to focus on other, more challenging tasks that are required to develop, test, and launch new systems.

Not involving the operators is only one of the pitfalls. In the next section, we will briefly highlight some other potential pitfalls.

Avoiding pitfalls in AIOps

Besides not involving operators during the implementation of an AIOps platform, there are three major pitfalls that must be avoided:

- **Not enough data**: With only a limited dataset, the AIOps platform will not perform. The system can never be accurate without enough data. Findings, issues, events, and anomalies will not or will only occasionally be detected and that will make the system unreliable, generating false positives or even mitigating issues that it doesn't need to solve.

- **Wrong or irrelevant data**: This might be even worse than not having enough data. Wrong data will definitively lead to false outcomes and unexpected or unwanted actions. In order to learn in the correct way, AI needs enough and valid data.

- **Data silos**: AIOps platforms get their data from multiple systems. AIOps, however, needs this data to be available from one source to enable data analytics. Operational data, application performance data, data on connectivity, and security data: it needs to be collected and aggregated to one source to make the system fast and efficient at creating valuable datasets. In other words, the data silos need to be broken down.

Summarizing, AIOps needs enough data, relevant data, and for it to be stored in one repository so it can create valuable datasets for operational analytics. By now, you should realize that AIOps systems are complex, but there is a growing number of tools that can help you to get started. We will list some of these tools in the final section of this chapter.

Popular tools in AIOps

Lastly, we will provide some tools that will get you started with AIOps. The AIOps landscape has grown tremendously in recent years. Take a look at the Periodic Table of DevOps tools in the following figure: the category of AIOps was only added very recently, already listing a significant number of tools, such as Splunk, Datadog, Dynatrace, New Relic, and the Elastic ELK Stack:

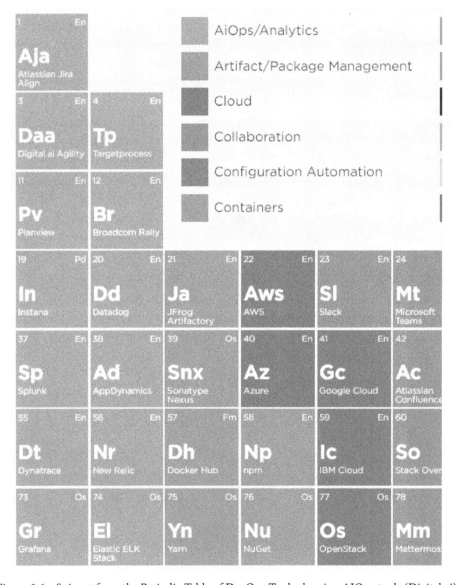

Figure 8.6 – Snippet from the Periodic Table of DevOps Tools showing AIOps tools (Digital.ai)

It's not easy to select the right tools. The reason for that is that these tools differ very much from each other. It depends on the angle that the enterprise takes to evaluate the proper tools and toolsets. Things to consider are the following:

- Tools that focus more on application and application performance management, rather than infrastructure components

- SaaS or on-premises tools

- Native AI and ML or tools that require extensions

- Data-driven AI or AI that is also context-aware

> **Note**
>
> Gartner publishes a magic quadrant every year to identify the leaders and promising newcomers in this emerging market. The report can be downloaded from Gartner's website at `https://www.gartner.com/en/documents/4000217/market-guide-for-aiops-platforms`. Be aware that you or your employer will need to become a client to be able to download content from the Gartner site.

Obviously, more sophisticated tools will come at a price. The most important question an enterprise must ask itself is what it wants to get out of AIOps. That will drive the business case and possibly justify the investment.

Summary

This chapter was a deep dive into AIOps. This is a rather new domain, but very promising. We've learned how AIOps platforms are built and learn as they are implemented in enterprises. It's important to understand that you need a logical architecture to have a complete overview of how systems fulfill functionality and how they are related to other systems, without already knowing the full technical details of these systems.

Next, we defined the key components of AIOps, being big data and machine or deep learning. AI only performs if it has access to enough relevant data on which it can execute analytic models. These models will teach the platform how to detect issues, anomalies, and other events, predict the impact on the IT landscape, find root causes faster, and eventually trigger actions. These actions can be automated. AIOps platforms will avoid a lot of tedious, repetitive work for operators, something that is called toil in SRE.

We've learned what the reference architecture of AIOps looks like and how an enterprise can successfully implement the system. In the last paragraph, we looked briefly at some popular tools in the field.

In the next chapter, we will learn how to integrate AIOps into DevOps.

Questions

1. What does the presentation layer in logical architecture do?

2. AIOps platforms are able to detect deviations from expected system behavior through algorithms. What do we call this process of detecting deviations, a key feature of AIOps?

3. AIOps works with operational system data and data coming from events such as incidents and problems. What is this latter type of data referred to by Gartner?

4. False or true: data cleansing is essential in AIOps.

Further reading

Pragmatic Enterprise Architecture, by James V. Luisi, 2014

9
Integrating AIOps in DevOps

So far, we've looked at automating development from a DevOps perspective and have automated operations. The next step is **artificial intelligence** (**AI**)-enabled DevOps. DevOps engineers manage multiple libraries and various pipelines. To speed up digital transformation, it's crucial that issues are detected and remediated fast. AI can also be of great added value in these DevOps processes. In this chapter, you will learn how to implement AI-enabled DevOps and enable rapid innovation.

After completing this chapter, you will have a good understanding of the various steps that need to be taken to implement and integrate AI-driven pipelines for development and deployment. You will be introduced to some major tools and will learn the requirements to implement these as part of the innovation of digitally transforming enterprises.

In this chapter, we're going to cover the following main topics:

- Introducing AI-enabled DevOps
- Enabling rapid innovation for digital transformation
- Monitoring pipelines with AIOps
- Assessing the enterprise readiness of AI-enabled DevOps

Introducing AI-enabled DevOps

In the previous chapter, we studied the AIOps platform, concluding that it will help operators in getting rid of tedious, repetitive tasks, detecting and solving issues faster, and enabling more stable systems. Stability and resilience are still the key aspects operators strive for with IT systems, yet new features and changes to the systems are being developed and launched at an increasing speed. If AI can help operations, it can also help development. This section will explain why AI-enabled DevOps will help in creating better systems at a higher velocity.

AI can help developers monitor and detect issues in their builds faster than if this were done only manually or even in an automated process, without the power of AI. With AI, it's possible to continuously monitor code changes, compare these to other code building blocks, and swiftly detect issues. But AI will also enable predictive mitigations: it will learn how certain code changes may impact systems. Given the fact that new features and thus new code are developed at an ever-increasing pace in systems that become more complex, AI-enabled DevOps is a solution to ensure the stability of these systems. AI will help in managing various code libraries, keeping track of configurations and deployment scripts, and avoiding unexpected application behavior.

How does it work? Well, in the same way as AIOps, which we discussed in *Chapter 8, Architecting AIOps.* AI-enabled DevOps will learn from patterns in the DevOps cycle. To do that, it needs data. It will use the code data that is stored in the code repository and the process data that is used to build the CI/CD pipeline, and then run the various test and deployment procedures. Additionally, it will learn from historical data; that is, issues and events that have occurred and the way these have been solved. Through analytics and machine learning, AI will learn how to optimize code builds, testing, and deployments.

DevOps has certainly evolved over the last decade, but developers and operators still face some tedious tasks in coding, testing, and deploying new features to systems. A lot of the work is still very manual and requires several steps in testing and code reviews. Since code is becoming increasingly complex and systems are becoming more entangled in various platforms, issues may not always be detected in time or at all. To save time, code review is sometimes done through sampling: some code is randomly picked, and that specific piece of code is reviewed, leaving no guarantee that the rest of the code is OK. So, when is code OK? Often, the focus is on removing empty lines or obsolete spaces, but these things can easily be solved with formatting tools. Obviously, bugs need to be removed, but code also needs to be optimized to perform well. All this needs to be done while production is kept running and stable.

AI, **machine learning** (**ML**), and deep learning can help overcome these issues. Then, there's AI-enabled DevOps. This requires the following components:

- Access and control over the source repository

- Data lake and data marts for modeling

- AI-integrated pipelines

The basic process contains the steps shown in the following diagram:

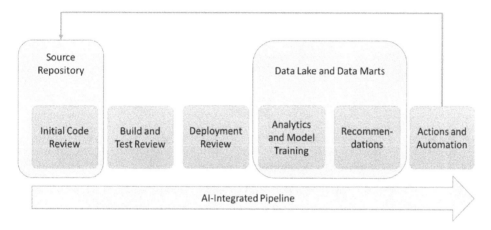

Figure 9.1 – Concepts of an AI-integrated pipeline

We will elaborate on these steps in the *Monitoring pipelines with AIOps* section, where we will discuss various tools and technologies for monitoring processes in DevOps pipelines.

AI is a new domain in DevOps. It's an innovation that can speed up the digital transformation of enterprises. Before we dive into the details of injecting AI into DevOps pipelines, it's good to get a better understanding of the innovation cycle and the rationale for including AI and ML. We will discuss this in the next section.

Enabling rapid innovation in digital transformation

The majority of modern enterprises are well underway in transforming their business to make it more digital native. We talked about this extensively in the first two chapters of this book. Customers continuously demand new features, and they want these features to be delivered almost instantly. To control this process, enterprises need to develop an innovation strategy catering for rapid innovation. An innovation strategy can be depicted as a pyramid, where AI-driven innovation is at the very peak of this pyramid.

This can be seen in the following diagram:

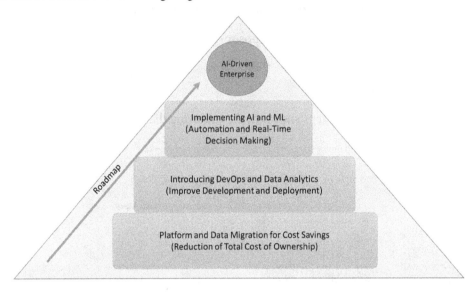

Figure 9.2 – Pyramid of AI-enabled innovation

Enterprises do not get to the top of the pyramid in one go; they usually start at the bottom, where innovation is driven by cost savings. From there, they need to develop the next steps, resulting in rapid innovation using AI and ML.

The first steps typically involve ways to find a budget, which they do by implementing technology that can lead to drastic cost savings, for instance, by moving to another platform. That choice of platform is crucial: enterprises will want to make sure that they make a sustainable choice for the future and have planned such innovations to drive digital transformation. Such a platform can be a public cloud, offering not just hosting services, but allowing us to use cloud-native technology and integrate that with DevOps, AI, and ML.

An even more important step is to collect and leverage the use of data. AI and ML need data, and it needs to be aggregated for analytics purposes and for training data models. Data is likely the most valuable asset in any enterprise, so this step will take a lot of time.

Aggregating data doesn't mean that every single piece of data needs to be in one data lake. There will be a need for different datasets, but architects will have to think of efficient ways to create these datasets without the usual silos in an enterprise. And then there's data security: who is authorized to see what data and for what reason? How do you prevent data from getting somewhere it shouldn't be? Data **identity and access management (IAM)** and **data loss prevention (DLP)** are important topics. For this, an enterprise will need to have a consistent system for data classification.

Now, we can bring AI and ML to the table. Data models need to be trained, deployed, and integrated with the CI/CD pipelines to enhance coding and application development. The pipelines will have to be integrated with the data analytics models and AI services from the platform of choice. The end of this journey might be an AI-controlled CI/CD pipeline. That's the topic of the next section, where we will have a look at some AI-driven tools in the major public clouds; that is, Google Cloud Platform, AWS, and Azure.

Monitoring pipelines with AIOps

In this section, we will study AI-driven technology that will help developers in monitoring and improving their CI/CD pipelines. Let's recap on the principle of a pipeline first. A pipeline should be seen as a workflow: it guides code through a process where it's tested and eventually deployed to a platform. Following this process, code will be pushed to different levels in the promotion path: development, testing, acceptance, and production. This process can be automated.

At the start of this process, and thus the pipeline, there is a repository where the various components of systems are stored. Since everything is code, the repository will hold code for applications, infrastructure components, configuration templates, and scripts to launch APIs. While building a system through a pipeline, DevOps software will make sure that the appropriate components are pulled from the repository and compiled into packages that can be deployed. A common way to do this is by using containers, such as Docker images that have been orchestrated via Kubernetes. Containers are also very suitable for injecting AI and ML into pipelines. The following diagram shows the basic functionality of Kubernetes:

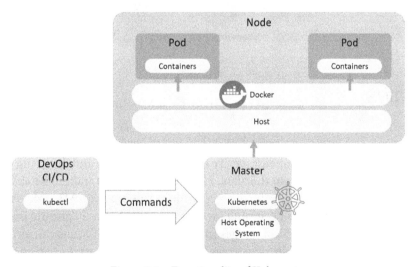

Figure 9.3 – Functionality of Kubernetes

From the DevOps CI/CD pipeline, Kubernetes can be instructed on how components must be pushed to the target platform using containers. To enable this, we can use the Kubernetes `kubectl` command-line tool. It runs commands against Kubernetes clusters and tells these clusters how to deploy application code while monitoring the cluster's resources and hosts.

Introducing Kubeflow by Google

According to their own documentation, Kubeflow is the ML toolkit for Kubernetes. It allows us to implement and monitor complex workflows and application development processes in pipelines that are running on Kubernetes and using ML on any cloud, such as AWS, Azure, and Google Cloud, just to name a few.

In-depth monitoring and analytics for all the components in the pipelines are highly valued features of Kubeflow. Originally, TensorFlow and PyTorch were used to train the data models. In the words of Kubeflow itself, it takes care of all the *boring stuff* by using ML, so that developers can concentrate on the new features.

The high-level architecture of Kubeflow is shown in the following diagram:

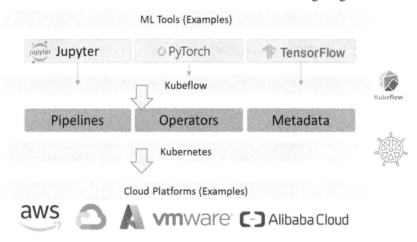

Figure 9.4 – Conceptual architecture of Kubeflow

ML is used to load vast amounts of data, verify that data, process it, and, by doing that, train models so that they learn from this data. However, developers want to be able to do this at scale. An example could be image recognition. You can train ML models to recognize images, a feature that is becoming increasingly popular in diagnostic imaging at hospitals. Using AI and ML clinical images from; for example, CT scans can already be valued, denoised, and have possible focus areas highlighted to support doctors in getting a more precise diagnosis faster. To run this type of model at scale, using containers is a viable option. This is what Kubeflow does.

> **Note**
>
> On https://www.kubeflow.org/docs/examples/, you can find tutorials and examples of Kubeflow use cases, including a sample for image recognition and a tutorial on how to use a Jupyter Notebook on a Kubeflow cluster.

The architecture of Kubeflow shows that AI-enabled DevOps requires a number of components and that different tools need to be integrated. Some of these tools are platform native, such as CodeGuru by AWS and MLOps in Azure. We will briefly evaluate these in the following sections.

Introducing CodeGuru by AWS

There are more developments underway, proving that this is a growing market. As an example, AWS announced the introduction of ML in DevOps in May 2021, using CodeGuru.

CodeGuru Reviewer is a tool that uses ML to detect issues and bugs in code, but also vulnerabilities in security policies that have been applied to code. It also provides recommendations to improve the code, either by solving bugs or suggesting enhancements to be made to the code.

A second component of CodeGuru is CodeGuru Profiler. Once the code review has been completed, Profiler validates the runtime of the code, identifying and removing inefficiencies in the code and, with that, improving the performance of the application. AWS claims that it also helps in decreasing compute costs, since Profiler also checks the code against the resources that it calls during the runtime process. By using ML, it can do this proactively, since it learns how code can be optimized, from comparing new code to existing code patterns. According to the AWS documentation, CodeGuru has already reviewed over 200 million lines of code since its launch.

The following diagram shows the architecture of CodeGuru using CodeGuru Reviewer and CodeGuru Profiler:

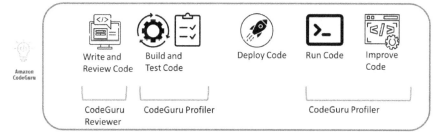

Figure 9.5 – CodeGuru and CodeGuru Profiler by AWS

The next step in this domain is Amazon DevOps Guru. DevOps Guru works more on the infrastructure level of DevOps. It runs pre-trained ML models that analyze system logs and metrics against operational baselines to detect anomalies in infrastructure components. Because it uses deep learning, the model is trained and enhanced continuously.

Introducing MLOps in Azure

All major cloud providers have AI-driven solutions. The last one we will discuss briefly is MLOps in Azure. The basic principle is the same as with CodeGuru: MLOps will pull code from the repository, which is usually integrated with Azure DevOps. The following diagram shows the MLOps architecture:

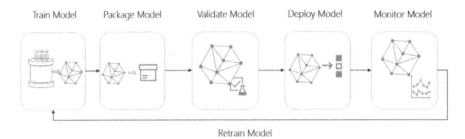

Figure 9.6 – Architecture of MLOps in Microsoft Azure

MLOps performs various tests on the committed code. Although the outcomes will be comparable to other AI-enabled tools, MLOps works a bit differently, using an Azure ML pipeline where the models are trained. However, MLOps also works on the foundation of containerization with **Azure Kubernetes Services (AKS)** and **Azure Container Instances (ACI)**.

In summary, using AI-enabled DevOps will result in you being able to do the following:

- Identify missing code.
- Detect badly written code.
- Detect unnecessary code.
- Detect expected and/or required but missing dependencies.
- Perform configuration checks.
- Make recommendations for improvements.
- Trigger automated actions for improvements.

AI-enabled DevOps is a fast-growing market, so besides the native services of major cloud providers such as CodeGuru and Kubeflow, which were launched by Google, there are a lot of emerging tools from startup companies that have been launched over the past few years. Examples include Pipeline.ai, Enterprise AI by DataRobot, Hydrosphere.io, and Xpanse AI. The first three focus more on ML-driven pipelines, while Xpanse AI is an environment for creating data models using AI.

> **Note**
>
> In this chapter, we're discussing DevOps pipelines that have been enhanced with AI and ML. To get started with data modeling and data analytics, you will also need tools to enable them. Popular tools in this field include Databricks, MapR, Cloudera, and RapidMiner.

To conclude, it makes sense for enterprises to invest in AI-driven DevOps, enabling rapid innovation and speeding up digital transformation. So, the first question that needs to be answered is, when is the enterprise ready for this paradigm shift? We will briefly discuss this in the following section.

Assessing the enterprise readiness of AI-enabled DevOps

So far, we've learned that digital transformation is a process. It doesn't come in one go; the enterprise needs to be prepared for this. It includes adopting cloud platforms and cloud-native technology. Enterprises will have legacy systems and likely a lot of data sitting in different silos, leaving the enterprise with the challenge that this data is used in an optimized way. It's a misperception to think that AI-enabled tools and data science can solve this issue from the beginning.

The enterprise will need to have a complete overview of all its assets, but also its skills and capabilities. First, data specialists will need to assess the locations, formats, and usability of data sources. The data scientists then will have to design data models. They can't do this in isolation: they will have to collaborate with DevOps engineers and the application owners to agree on things such as version control, model training, and testing.

The agreed-upon data models can then be integrated into DevOps pipelines. In the previous section, we learned that tools will have to be selected and integrated with the CI/CD tooling, for example, by using containers and container orchestration platforms such as Kubernetes. Engineers specialized in ML can help in implementing, tracking, and training these data models by leveraging the benefits of AI.

That's not all. Processes – we called these engagement processes – such as incident, problem, and change management need to be aligned with the new development and deployment setup. Service managers will need to understand the new metrics that come with ML- and AI-enabled platforms. What needs to be done in the case of alerts or recommendations coming from the platforms? What is monitored and for what reason? Can tasks be automated even further, saving time and thus costs?

However, the most important question is, what recommendations need to be actioned to improve development and speed up releases? AI can help, but it needs time to *learn* the enterprise. Trained and skilled staff are required to help AI become familiar with the enterprise. It isn't magic.

In summary, the enterprise needs to have the following:

- Full visibility of all the assets
- Full visibility of all the data sources and how these are related to business processes
- Engagement processes that are implemented throughout the enterprise for consistency
- Trained and skilled staff, such as DevOps engineers, data scientists, and AI/ML engineers
- An innovation roadmap with realistic timelines

The aforementioned roadmap may lead to a distant horizon where operations are completely automated by means of AI and ML and no manual intervention is required anymore. That's where NoOps comes in. In *Chapter 10*, *Making the Final Step to NoOps*, we will introduce this concept.

Summary

In this chapter, we learned how to integrate AI and ML into our DevOps pipelines. We discussed the basic requirements and steps for implementing AI-enabled DevOps, starting with access to source repositories, creating data lakes, initiating and training data models, and follow-up recommendations and actions. We also learned that AI-enabled DevOps is a stage in digital transformation, but that enterprises need to set out a roadmap that eventually allows them to integrate AI and ML into their development and deployment processes. AI-driven development and operations are at the peak of innovation in digital transformation.

Next, we introduced some tools that will help us in implementing AI-enabled DevOps. We learned that it's a fast-growing market where major cloud providers try to integrate their native DevOps tools with AI and ML. Examples include Kubeflow by Google, CodeGuru by AWS, and MLOps by Microsoft Azure.

Finally, we discussed the readiness assessment for enterprises that want to implement AI-enabled DevOps. It's crucial to develop a comprehensive roadmap, including the different steps and a realistic timeline. The end of that roadmap might be a fully automated pipeline orchestration without any manual intervention: NoOps. This is the topic of the next chapter.

Questions

1. We introduced the innovation pyramid for digital transformation. What is the base platform of this pyramid?

2. Integrating AI into DevOps pipelines is typically done through containerization. We discussed a container orchestration tool that allows us to agnostically deploy containers to various platforms. What is this tool called?

3. Name three possible outcomes/results of AI-enabled DevOps, specifically for improving code.

Further reading

- *Pragmatic Enterprise Architecture*, by James V. Luisi, 2014

- Documentation on Kubeflow: `https://www.kubeflow.org/`

- Blog about the introduction of CodeGuru, by AWS: `https://www.allthingsdistributed.com/2021/05/devops-powered-by-machine-learning.html`

- Blog on MLOps in Microsoft Azure, by Lee Stott: `https://techcommunity.microsoft.com/t5/educator-developer-blog/machine-learning-devops-mlops-with-azure-ml/ba-p/742150`

10
Making the Final Step to NoOps

Is it possible to execute IT operations without hands-on operations? Research and advisory companies, such as Gartner and Forrester, foresee an IT future based on NoOps. The big idea behind NoOps is that literally everything can be automated. It means an even bigger role for AI, and something called **heuristic automation**. How can an enterprise move to NoOps, and what is the role of an architect in this field? We will discuss this in this chapter.

After completing this chapter, you will be able to explain NoOps as a concept and why enterprises should adopt the principles of NoOps. You will learn what heuristic automation is and how it's driving the architecture of NoOps. The most important lesson that you will learn is that NoOps is not simply about having no need for any operations at all.

In this chapter, we're going to cover the following main topics:

- Understanding the paradigm shift to NoOps
- Understanding the role of AI in NoOps
- Creating an architecture for heuristic automation
- Defining the roadmap to NoOps

Understanding the paradigm shift to NoOps

In the previous chapters, we discussed the introduction of **artificial intelligence** (AI) and **machine learning** (ML) into operations and development. In *Chapter 9, Integrating AIOps in DevOps*, we learned how an enterprise can leverage AI and ML in DevOps pipelines. The reason to do this is to make a lot of manual tasks obsolete through intelligent automation. NoOps takes all of this one step further: automate IT systems completely so there's no need for operators to manually intervene in the systems. How far away are we from that paradigm shift? In addition, *is it realistic?* We will discuss that in this section.

To answer the last question: NoOps seems to be more of an ideal than a real practice. The discussion around NoOps was initiated through the idea that teams could actually automate a lot of processes in development, especially regarding the deployment of applications. This started with services being provided as **Software as a Service** (**SaaS**) propositions, meaning an enterprise didn't need to worry about maintaining the services anymore. Updates and upgrades in SaaS and **Platform as a Service** (**PaaS**) were taken care of by the provider. Getting software and maintaining applications became as easy as working with a smartphone: at night you put your phone away, the manufacturer of the phone updates your phone, and when the phone is started again in the morning, all of the applications still work. In reality, for this to happen, a lot of actions must be executed in the background. In other words, NoOps might be superficial.

However, the idea of NoOps does deserve some credit. It's in line with the principles of DevOps and automating as much as possible in the development, deployment, and operation of code. According to Deloitte, NoOps was a logical evolution in automating IT tasks and shifting activities from operations to development, focusing on business outcomes. That's nothing new. DevOps and especially **Site Reliability Engineering** (**SRE**) have these as basic principles.

To be clear, NoOps is not the same as automated infrastructure provisioning. NoOps is about automated management of the full stack – applications, middleware, databases, and infrastructure. In terms of infrastructure, the concept of NoOps relies on infrastructure components that can be coded and controlled by code. Virtual machines might be an option, but these are still not very flexible in terms of automation of operations. Containers, and especially serverless solutions, are more logical.

We can automate coding, provisioning of infrastructure, deployment of APIs, and configurations. We can add technology that will detect issues and anomalies fast, and maybe even have systems automatically remediate these based on predefined policies. We might be able to automate one application completely, but the reality is that today's enterprise IT consists of complex ecosystems, within and outside the enterprise. This will make it hard, if not impossible, to really have NoOps in practice. On the contrary, operations are getting more complex. But then again, NoOps is not about getting rid of operations completely.

NoOps should be perceived as a concept and guidance in leveraging automation, saving costs, and speeding up developments, while also keeping systems stable, resilient, and unhindered by manual operational actions. NoOps will help architects with shift-left principles, injecting AI into operations, and also help in automating IT. We will explore shift-left and AI-enabled operations further in the next sections.

Understanding the role of AI in NoOps

In the previous section, we discussed whether NoOps provides a realistic way of operating for future enterprises. We've concluded that NoOps should be seen as both a concept and a way of thinking to leverage automation for operations. It's not about getting rid of operations as a whole – IT environments in enterprises have become too complex for that. Still, they have the challenge of using their IT talents in the most optimized way.

IT talent is becoming scarce, since the market demand for skilled and trained engineers is increasing at high speed. Because of this scarcity, the costs of staff are also increasing. To keep costs down and still be able to work as agile as possible, enterprise architects will have to search for other ways to operate IT. IT talent can then fully focus on developments.

However, operations will be needed. We need people to look after systems, and make sure that these systems are running stably. You can't leave it all to machines. For that reason, it's good to narrow the definition of NoOps down a little bit. We can define NoOps as the stage where dedicated operators are no longer required to manage IT. It would then be a logical step in the evolution of enterprise DevOps, where teams operate as one in development, deployment, and operations.

There are two things that will play an important role to get this stage:

- **Adopting shift-left**: In *Chapter 7, Understanding the Impact of AI on DevOps*, we concluded that shift-left is also applicable to both deployment and operations in DevOps. With automated templates, pre-approved patterns, and processes, we can start testing in an early stage of development and deploy consistent, stable code to the next stages (including production). It will certainly lead to less work for operations.

  Operators will typically define how production is monitored, whereas developers control this themselves in *sandboxes*, development, and testing environments. In NoOps, this is shifted to the very beginning of the whole cycle – the DevOps teams decide how code is tested and monitored. The team is end-to-end responsible. The distinction between operators and developers becomes obsolete.

- **Artificial intelligence and machine learning**: As we've seen with AIOps, AI and ML will help in detecting issues quickly, and can even be trained to give recommendations or initiate actions such as these:

  - **Automated software life cycle management**: AI will help in recognizing when software needs to be updated and will take care of the update process, keeping the services that run on the software operational and stable. Think of the smartphone analogy in the first section. The phone is updated, and when the user switches it back on, all functionality will normally be there again. This is enabled by the fact that the phone *knows* the dependencies in different apps, along with the underlying protocols and code. It uses AI for this.

  - **Automated remediation**: AI can trigger automated actions to fix issues even before the issue materializes. Predictive maintenance is enabled.

  - **Auto-healing**: When an unexpected issue occurs, AI will detect it, know how to solve it from history or by learning, and eventually apply the fix. The process of solving the problem is referred to as auto-healing. Be aware that all these actions – life cycle management, remediation, and healing – need to be logged so every change is traceable.

Now, we have discussed automation in DevOps, introduced AI in DevOps, and also discussed leveraging automation even further in NoOps. The foundation under this level of automation is heuristic automation. In the next section, we will study architecture for heuristic automation.

Creating an architecture for heuristic automation

First, let's get a definition of heuristic: in the literature, it is referred to as applying a solution to an issue without the aim of being the optimal solution, but sufficient to fix the immediate problem that was discovered. *Trial and error* would certainly match this definition. The Hungarian mathematician George Pólya used the term in his book, *How to Solve It*, first published in 1945. He provided some practical ways of solving problems.

One of his principles is commonly used in architecture applying ML: if you don't have a solution, assume that you have a solution and see what it does. Keep the good stuff and analyze the bits that didn't work well. Try the iterated solution again and learn from it. This is the base of heuristic automation. It uses heuristic learning that can be leveraged through AI that is able to recognize and learn from patterns. AI will use algorithms and automation – it constantly learns and adapts the analytics, up until the point where the solution is optimized.

First, we will need to understand the principle of learning. AI and ML in automation typically use deduction. This is shown in the following figure. There are more learning principles, such as induction and abduction, but these do not really add value to the goal of automation:

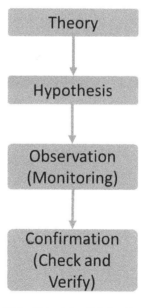

Figure 10.1 – Principle of deduction in ML

With deduction, the system will analyze an event by observing and comparing it with previous experiences, examples, and common theories. Based on that analysis, it will reach a conclusion. A rule base will tell the system what to do next. Obviously, this rule base is dynamic, since the system will add things it has learned to the rule base.

With that in mind, we can define the components for heuristic automation:

- **Event sources**: Aggregating data from events throughout systems and system components.

- **Data on patterns and correlations**: Repository with known patterns and correlations between systems.

- **Analytics engine**: Algorithms and rules enabling analytics and the prediction of outcomes.

- **Analytics processes**: The processes that tell the system what to do if anomalies are detected, and what solutions should be applied from the rule base.

- **Dynamic rule base**: Possible solutions based on induction patterns. This is a dynamic process, since solutions will continuously be improved. Based on the findings, the solutions are updated.

The logical architecture for heuristic automation is shown in the following figure:

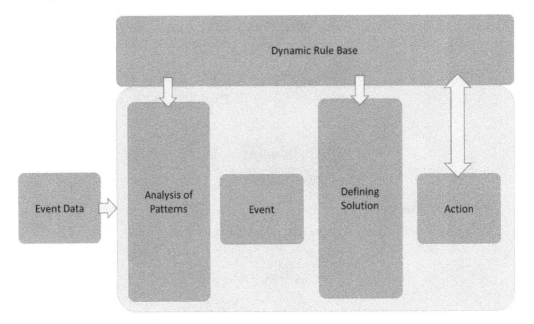

Figure 10.2 – Logical architecture for heuristic automation

Heuristic automation will enable the enterprise to do the following:

- Identify and solve issues without human intervention through the automated analysis of complex systems.

- Recognize, identify, and understand relationships between systems and predict the impact of issues.

- Solve issues through ML using various data sources and real-time analytics.

We stress once more that NoOps and heuristic automation by no means intends to replace operators. It will help them in solving issues faster, as well as helping businesses with more stable and resilient systems.

Defining the roadmap to NoOps

NoOps is not a leap of faith. As with DevOps and AIOps, any next step requires a plan and a roadmap. Using the principles of heuristic automation and AIOps, we can leverage automation for intelligent automation, collaborating with cloud-native automation, and automated application deployment in CI/CD pipelines.

The following figure shows how NoOps consists of three major components. These three components are the roadmap to get to a level of automation where dedicated operations are no longer required. DevOps teams are end-to-end responsible for the development, deployment, running, and maintenance of the code, supported by fully automated processes and AI:

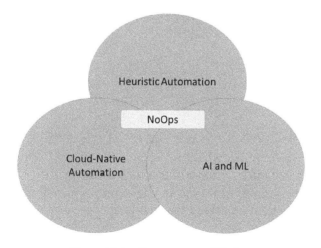

Figure 10.3 – Components of NoOps

The end state is predictive automation using predictive analytics that analyzes current data to eventually make predictions for the future. It includes the following:

- Scaling for future needs, for instance by analyzing usage of software, predicting future use, and actioning this in scaling the systems

- Foreseeing business trends and preparing systems for this, including preparing code by collecting and analyzing requirements

- Code suggestions for improving code, fulfilling predicted future needs

A very simplified roadmap to NoOps could look like the following figure:

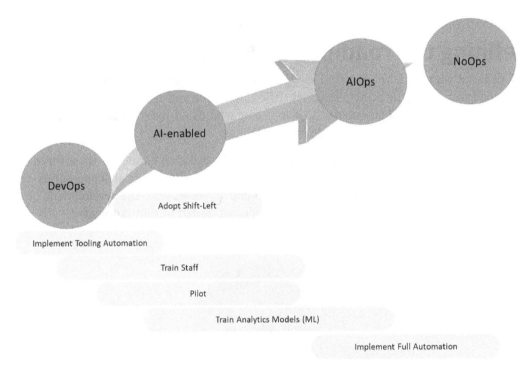

Figure 10.4 – Simplified representation of NoOps roadmap

Adopting NoOps requires both the right mindset and full support from the enterprise's management. Management will need to learn to trust automation. Only when this trust is established will they be willing to invest in the right tools. This will take time, and that timeline needs to be realistic, as reflected in the roadmap. Lastly, be prepared for pushback and setbacks. NoOps and full automation are more about adoption than technology, and adoption is a process.

Is NoOps a leap of faith? Could there be automation without human intervention? Think of this: already in 2018, the Open Road project has been defined, studying the possibilities for computers to design their own new generation of chips. In June 2021, the science magazine Nature published an article stating that computers no longer need humans to design new chips. Using AI, computers can design chips in a matter of hours, whereas humans would need months for the same outcome. The foundation technology that was used for this was ML, and the use of cloud-native analytics tools in Google Cloud to run parallel sessions and achieve predictability in the outcomes of the design.

In this part of the book, we extensively discussed the approach for DevOps, SRE, and AIOps. All of these methodologies leverage automation in development and operations, to the point that human intervention might become obsolete. The subsequent question, then – is this safe? Can we really leave IT to automation and AI? How secure will our systems then be? We need to map our DevOps strategy to security standards, including industry security frameworks and enterprise proprietary security policies. These need to be embedded in DevOps. This is the domain of DevSecOps. In the third part of this book, we will discuss this further.

Summary

In this chapter, we discussed the concept of NoOps – no operations. We discovered that the term might be misleading, since NoOps doesn't mean that the enterprise will no longer need operations at all. NoOps is a concept that leverages automation to its maximum potential. By automating development, deployment, and operations, scarce IT talent can focus on new features, since NoOps will help them by identifying and solving issues in IT systems fast. We've learned that NoOps (like AIOps) uses AI and ML. But NoOps also means that enterprises will need to embrace the shift-left mentality.

We've also learned that NoOps requires a specific type of architecture for heuristic automation: applying iterative solutions, learning from these solutions, and continuously improving them. We also discussed the different components of heuristic automation. In the final section, we explored a possible roadmap from DevOps to NoOps. We concluded that we already have the technology available, but that enterprises will need to adopt it to really leverage full automation and, with that, the concept of NoOps.

In the next part of this book, we will examine an important topic related to enterprise DevOps – security. We will learn how to integrate security with DevOps in DevSecOps.

Questions

1. A system detects an unexpected issue, knows how to solve it from history or by learning, and eventually applies the fix automatically. What do we call this type of action?

2. What type of learning is typically used by AI-driven systems to automate remediating actions?

3. True or false: predictive automation is applied to scaling that predicts the future usage of systems.

Further reading

-   Blog about NoOps on CIO.com by Mary K. Pratt: `https://www.cio.com/article/3407714/what-is-noops-the-quest-for-fully-automated-it-operations.html`

-   Blog about heuristic automation of infrastructure by Ramkumar Balasubramanian: `https://www.linkedin.com/pulse/heuristic-automation-infrastructure-ramkumar-balasubramanian/`

Section 3:
Bridging Security
with DevSecOps

In the previous chapters, we automated development and operations. Now, the final step is to automate security along with it. This is where DevSecOps starts: it integrates security in every step of development and the release to operations. DevSecOps includes testing, deployment, and final delivery. In this part, you will gain a deep understanding of DevSecOps and learn what the architecture looks like and how security must be integrated with DevOps.

The following chapters will be covered under this section:

- *Chapter 11, Understanding Security in DevOps*

- *Chapter 12, Architecting for DevSecOps*

- *Chapter 13, Working with DevSecOps Using Industry Security Frameworks*

- *Chapter 14, Integrating DevSecOps with DevOps*

- *Chapter 15, Implementing Zero Trust Architecture*

11
Understanding Security in DevOps

You can't talk about the cloud, modern apps, and—for that matter—digital transformation without talking about security. A popular term is *security by design*. But even security by design needs to be embedded in the enterprise architecture. It also applies to the DevOps cycle: DevOps needs to have *security by design*. Before we can discuss this and principles such as zero-trust, we need to get a good understanding of security first and how it's impacting the DevOps practice. This chapter provides an introduction to security in DevOps.

After completing this chapter, you will have learned why it's important to include security in the enterprise architecture and how an architect can collect and assess risks, and be able to identify what specific risks are in DevOps. You will also learn about setting security controls and what the main topics are that need to be addressed in DevSecOps.

In this chapter, we're going to cover the following main topics:

- Embedding security in enterprise architecture
- Understanding security risks in DevOps
- Getting DevSecOps-savvy
- Defining requirements and metrics

Embedding security in enterprise architecture

It's a topic you can read about practically every day: businesses that have been hit by some sort of hack or attack. The smallest hole in a system will be found by criminals and exploited. Currently (in 2021), the most *popular* attacks are the following:

- Ransomware
- Phishing
- Denial of service

The first two, ransomware and phishing, really exploit holes in the defense layer of enterprises. The last is basically about bombing a system so heavily with traffic that the system eventually collapses. All three are fairly easy to execute. In fact, you can buy software and even services that will launch an attack on the targeted address. And no, you don't have to go to the dark web for that. It's out there, in the open, on the *normal* internet.

How can an enterprise protect itself from these attacks? First of all, it's important to realize that the IT of any enterprise has become more complex, as we have seen in previous chapters. IT systems are no longer only in one privately owned data center, but it has become an ecosystem using the public cloud, private stacks in privately owned data centers or colocations, **Platform as a Service (PaaS)**, and **Software as a Service (SaaS)**. Security must be intrinsic in every service that an enterprise uses.

The four traditional principles of enterprise security are as follows:

- Prevention
- Detection
- Correction
- Direction

The first three are about detecting security issues, correcting them with mitigating actions, but obviously, it's better if issues can be prevented—hence, prevention is priority one. Direction is about guidelines and guardrails: enterprises defining policies and security standards to keep all systems secure. That's where the fifth principle comes in: consistency. In an enterprise with various divisions, clusters, and teams, you need to ensure that security is implemented in every corner of that enterprise. You can't have one division or team not adhering to the enterprise's security. In other words: the strength of the chain is defined by the weakest link.

So, where do we start with enterprise security? Enterprise architecture frameworks can help to get you started. An example is **Sherwood Applied Business Security Architecture (SABSA)**, which offers a methodology to define security architecture. It consists of six layers, as shown in the following figure:

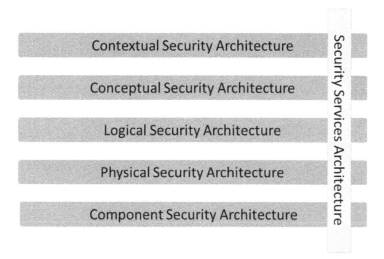

Figure 11.1 – The SABSA model for security architecture

Information Systems Audit and Control Association (ISACA) combines SABSA with the generally more known **Control Objectives for Information and Related Technology (COBIT)** five principles. These five principles are as follows:

- Meeting the needs of the stakeholders of the enterprise
- Covering the enterprise end to end
- Applying a single integrated framework
- Enabling a holistic approach
- Separating governance from management

The combination of SABSA and COBIT leads to a top-down approach to define the architecture for the entire enterprise. The enterprise architect will have to execute at a minimum the following steps, in cooperation with the chief security architect:

1. Identify the business goals: This is the first stage in enterprise architecture. Enterprise architecture always starts with the business strategy, goals, and objectives.
2. Identify the business risks.
3. Identify the required controls to manage risks.

4. Design and implement these controls, for instance, the following:

    - Security governance processes

    - Access controls

    - Incident management processes

    - Certificate management processes

5. Design physical architecture for, among others, used platforms, networks, operating systems, and datastores. The physical architecture should also include the cloud platforms and services such as PaaS and SaaS and DevOps practices such as CI/CD.

6. Validate business and physical architecture to the compliancy and security standards and protocols the enterprise must adhere to.

7. Define and implement operation architecture, for example, the following:

    - Configuration management

    - Monitoring

    - Logging

    - Change management

> **Note**
>
> The preceding provided list is not meant to be exhaustive. In the *Further reading* section, we included a link to the ISACA journal about enterprise security architecture. In that journal, you will find more detail on the steps as described, using SABSA and COBIT.

An important topic in defining the security architecture is understanding risks. What are the risks, which systems are at risk, and what is the impact on the enterprise and its business? DevOps brings along its own risks. We will discuss this in the next section.

Understanding security risks in DevOps

There's a classic cartoon on the internet. It shows a boxing ring. The speaker announces an immense set of security tools and rules in the left corner of the ring. Then, in the right corner, he announces Dave: a nerdy-looking guy, wearing a shirt saying *human error*. The message: you can have every security system in the world, but it won't stop human error. And development is still mainly work done by humans. Humans make mistakes. Is that the biggest risk in DevOps or are there other specific risks that need attention? We will discuss this in this section.

To answer the question of whether DevOps implies specific risks, yes. Implementing DevOps without paying attention to security will definitely increase the risk of attacks, simply by raising the attack surface of systems. There are three main topics that need to be addressed:

- **Access management**: DevOps teams likely use code repositories that are manually accessed either by developers or by tools. Code needs to be protected, even when operating in open source mode, and code is shared so that more developers can contribute to the code. Even in that case, companies would want to regulate access to code so that it doesn't get out into the open, or worse, malicious code is injected.

  You need a role-based access model to the code repositories: who has read rights and write rights and who has full access—and for what reason?

  Keep track of the accounts. For example, GitHub companies can have internal repositories that only assigned staff—or tools—of that company can have access to. Within that internal repository, the administrator delegates the roles. Credentials are set according to the security policies of the company. A recommended way to keep access control is to implement **privileged access management (PAM)**.

  Because of DevOps, teams will have to create more privileged accounts that are shared—either manually or automatically—among developers and tools. These accounts also involve service accounts, encryption keys, SSH keys, and API certificates that are all kept in the repositories. Unauthorized access to these repositories is disastrous.

  A PAM solution offers a way to authorize and audit any activity in the DevOps cycle. Most of these solutions use key vaults to keep access details secured and to authenticate and authorize users before they can actually access repositories.

- **Missing guardrails and guidelines for a DevOps way of working and tools**: Access to code is one thing; the next thing is: what do we do with that code? It's very unlikely that enterprises would allow DevOps teams to just commit and push new code to production. First of all, the enterprise or chief architect—in most larger enterprises, there will be a group of leading architects—really needs to think of the preferred toolset.

  It's important that every DevOps team uses that toolset instead of implementing tools of their own choice, however tempting it might be. The issue is that DevOps toolsets do not have a common standard for security policies, such as access control. That is something that the enterprise itself needs to establish and implement, which is easier if you go with one toolset.

The same applies to the way of working: that needs to be consistent throughout the entire enterprise; we can't stress that enough. All of this has to be defined in a preferred technology list and DevOps guardrails and principles. Typically, there will be a master branch. New code will first be pushed to a separate branch—often referred to as the feature branch—where it's tested. After the validated, positive test results, the code is merged to the master. The master code or main branch is again stored in the repository. That principle is shown in the following figure:

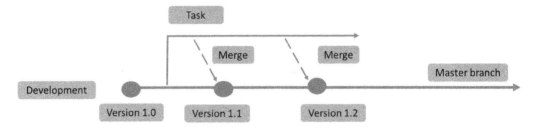

Figure 11.2 – Principle of merging new code

The guardrails define when code is pulled from the repository, how it's committed to feature branches, and how it's tested and eventually released for merging with the master.

- **Focus on the development process and velocity, neglecting security**: DevOps is all about gaining speed in development. That should never be an excuse to neglect security. In the following section, we will illustrate that with a real-life example.

To summarize, DevOps security is about three topics:

- **Traceability**: Track every action in the DevOps cycle and pipelines.

- **Auditability**: Ensure that systems that are developed in DevOps are compliant with the security standards of the enterprise and the industry frameworks that the enterprise is submitted to. *Chapter 13, Working with DevSecOps Using Industry Security Frameworks*, talks about that in more detail.

- **Visibility**: Have solid monitoring systems in place.

Now we have a good understanding of the security risks in DevOps. In the next section, we will discuss how to get started with DevSecOps.

Getting DevSecOps-savvy

Security starts with access to the repositories, the source of code where the DevOps cycle begins. As we've learned so far, we want to automate as much as we can in development, testing, and deployment. Next, by adopting DevOps, businesses want to speed up the development and deployment of new applications and features. Speed and agility might lead to security risks, because code is not sufficiently tested or, worse, it's pushed to production without applying the proper security policies to gain time. Let's illustrate that with a real-life example.

Developers fork code from the repository and start working on that code. At a certain stage, it needs to be pushed to designated infrastructure to run that code. In development, the code runs fine, since it's not interfacing yet with production systems. As soon as the code is ready to release in production, it will need to establish those connections. Commonly, in enterprises, specific routing and firewall ports need to be opened to allow for that connectivity and transfer of traffic.

Firewall rules and, more specifically, opening ports and assigning firewall rules are typically not done automatically; security engineers will want to assess the requests before they approve for settings to be implemented. In a lot of cases, this is really slowing down the whole DevOps and agile process. But in almost every enterprise this is the practice: security being the last stop before final deployment and not embedded in the DevOps cycle. Bypassing this is, however, a bad idea. It will increase the attack surface of code and systems.

Bottom line: security must be embedded in DevOps—it makes development and deployment a shared responsibility of developers, operators, and security engineers. Security is not just the responsibility of the security team, but of the whole DevOps team.

Starting points for DevSecOps

Security scanning starts as soon as code is pulled from repositories. The first thing to do is to have a **role-based access control** (**RBAC**) model applied to the repository. RBAC defines who has access and to what level. Is an identity allowed to only view code, or is full access granted to pull code, change code, and add code to the repository? Be aware that this doesn't necessarily have to be a person. DevOps tools are *identities* too and you need to think of the required access to these tools.

Throughout the DevOps cycle, code is constantly reviewed and scanned and tested on vulnerabilities. A very simple example is if connectivity to the internet is prohibited from an application and the code shows a pointer to port 80, it might be flagged as a risk. It might be allowed if an application is relying on a service to which it connects through the internet, but then that needs to be verified against the security policies. Scanning, testing, and validation need to be done as early in the development process as possible. Also, here, the shift-left principle applies.

The greatest benefit of embedding security in DevOps is the fact that we can also have automation to do checks and even apply patches and fixes as soon as a vulnerability is discovered or acknowledged. For instance, if a new release of a code base contains patches for **Common Vulnerabilities and Exposures (CVEs)**, this can be injected into the security baseline and integrated with security testing procedures. Code is automatically checked against this new baseline. Software can execute tests, validate whether code is compliant, and if the code does not pass, flag in the case of issues or trigger automated remediation by installing the patches. This doesn't only apply to application code, but certainly also for the used infrastructure, operating systems, and middleware.

DevSecOps using containers

Another example: a growing number of enterprises work with containers for code distribution. Just like virtual machines, containers also need to be compliant with security policies. Enterprises will likely use hardened containers, setting policies for the following:

- **Hardened host operating systems**: Often, these are Linux operating systems that protect the host from breaches by infected containers. Examples are security packages that are applied to Linux hosts, such as SELinux and seccomp; these Linux distributions allow **Mandatory Access Control (MAC)** to the kernel settings. Enterprises can opt to develop their own **Linux Security Modules (LSMs)**.

- **Container runtime security policies**: Setting specific permissions for mounting containers, privileged containers, disabling SSH in containers, settings for binding ports, and exposing ports.

The following figure shows the principle of CVE-based security scanning for containers:

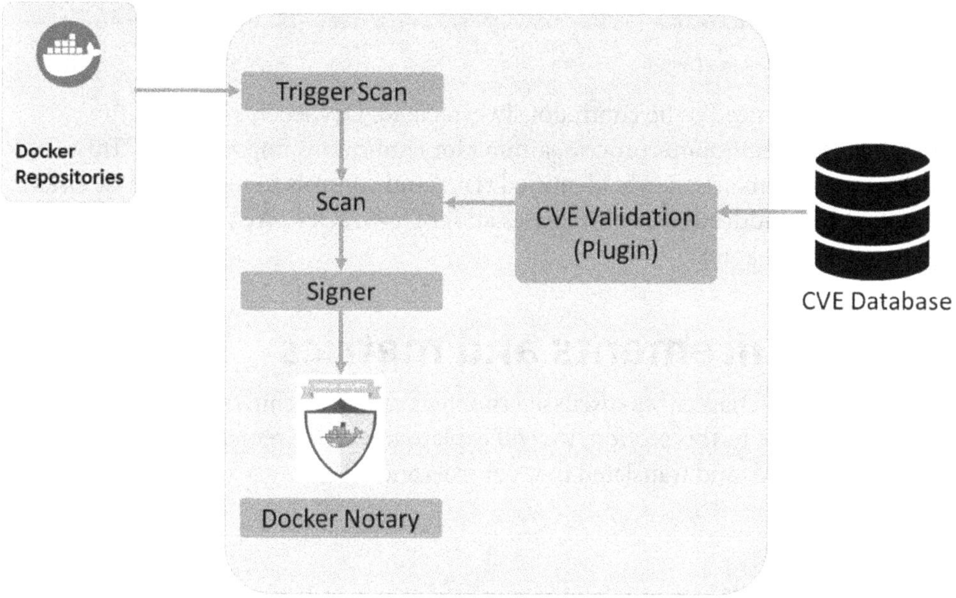

Figure 11.3 – Concept of Docker security scanning

The scan is done with the data from the CVE database. This can be data coming from, for example, **National Institute of Standards and Technology** (**NIST**), MITRE, or suppliers such as Microsoft who also issue CVE notifications for their products and services. After the scan, Docker *signs* the image that is pulled from the repository. For this, it uses Docker Notary, which verifies the image and prevents developers from using images that are not signed.

We now have hardened, validated containers with specific privileges set. Now, the next most important thing is to control the settings; once the privileges to a container are set, there should be no way to alter these privileges. Containers should not be able to gain new privileges; otherwise, they are not hardened. In Docker, there's a simple way to check and set this. Using the following command, you can list all the security settings for containers:

```
docker ps --quiet --all | xargs docker inspect
--format 'SecurityOpt={{ .HostConfig.SecurityOpt }}'
SecurityOpt=[label=disable]
```

It will return messages on hardening of the containers, creation of separate partitions for containers, and audit configuration for Docker files and directories.

Next, set the no new privileges option:

```
docker run <run-options> --security-opt=no-new-privileges
<image> <cmd>
```

Obviously, these settings need to be continuously evaluated. DevSecOps is—as with DevOps—a repeatable, continuous process, aiming for continuous improvement. This means that goals, requirements, newly identified risks, and controls to mitigate these risks need to be assessed and addressed as enterprises adopt DevSecOps. We will discuss this in the final section of this chapter.

Defining requirements and metrics

In the first section of this chapter, we discussed the steps that an architect must take to define enterprise security. In this section, we will explain how requirements and metrics can be collected, validated, and translated into controls and KPIs.

Business goals

We've talked about this in *Chapter 1, Defining the Reference Architecture for Enterprise DevOps*, but obviously, it's important to understand the goals a business wants to achieve. What markets are they in, how do they serve customers in these markets, and what is the product portfolio? It does make a huge difference if a business is operating in financial products or healthcare. Their markets define the risk level. The risk for a bank or an investment company could be mainly financial, whereas for healthcare, the biggest risk could be involving the life of patients. The goals will be different too: an investment company might have the goal to support as many businesses with funds, whereas healthcare companies will have the goal to cure people with specific solutions. So, the business and the goals will set the business attributes.

Business attributes

Attributes can be the availability of systems, the accuracy of data, and the privacy of customers. Next to the type of business and the goals, regulations will set levels of these attributes. Financial institutions will have to adhere to financial national and international regulations such as **Sarbanes-Oxley (SOx)** and healthcare providers to the **Health Insurance Portability and Accountability Act (HIPAA)**, as an example. Be aware that these regulations are audited. We will talk about this in *Chapter 13, Working with DevSecOps Using Industry Security Frameworks*.

Risks

Based on the business goals and the attributes, we can define the risks. What will be a major threat to the business if a certain event occurs? If you think about availability, it could be the situation that an enterprise doesn't have a means to execute a failover if crucial systems stop for any reason. And how can the enterprise recover from major outages? Not having redundant systems can be a risk, just like vulnerabilities in applications that can be exploited by criminals. The MITRE ATT&CK framework can help in identifying risks; it will also be discussed in *Chapter 13, Working with DevSecOps Using Industry Security Frameworks*.

Controls

The next step is to define risk controls. Examples can be the following:

- Create a business continuity plan that caters to disaster recovery.

- Implement a **public key infrastructure** (**PKI**) with identity stores and vaults to ensure the privacy of users.

- Implement application firewalls with specific firewall rules to protect critical systems.

- Set controls to manage, update, and upgrade all of the preceding: maintaining the business continuity plans, managing the security policies, and evaluating the firewall settings on a regular basis.

Validate that these controls are linked with the attributes and the applied security frameworks for auditing. For every control, there has to be a rationale that can be verified in audits.

The good news is that enterprises do not have to think about these controls all by themselves. The **Center of Internet Security** (**CIS**) has defined controls for a lot of IT domains: the CIS Controls framework. The basic CIS Controls include controlled use of privileged access, secure configuration of all IT system assets (including containers), and control of network ports, protocols, and services. For Azure, AWS, GCP, as well as for container platforms such as Kubernetes, CIS has defined specific controls.

CIS, MITRE, and other frameworks and how they impact DevOps will be further discussed in later chapters, but first, we will learn what a DevSecOps architecture should include. That's the topic for the next chapter.

Summary

This chapter provided an introduction to integrating security into DevOps, discussing the concept of DevSecOps. We've discussed the importance of security in enterprise architecture and how this is also driving security in enterprise DevOps. We've learned about the main security risks that are involved in adopting DevOps, and we had a closer look at securing containers as one of the most used technologies in DevOps practices. With that, we defined some critical starting points for adopting DevSecOps.

In the final section, we learned how to collect and assess risks from business goals and business attributes, introducing commonly used security controls frameworks such as the frameworks by CIS. With some examples, we explored the various steps that an architect needs to take to have a security standard that can also be applied to DevOps.

In the next chapter, we will explore the architecture of DevSecOps in more detail, before we start integrating security policies and industry frameworks with a DevOps way of working.

Questions

1. Name the four traditional principles of enterprise security.

2. What does Docker use to validate signed containers?

3. True or false: Security should not hinder speed and agility in DevOps.

Further reading

- ISACA journal about enterprise security architecture using SABSA and COBIT: `https://www.isaca.org/resources/isaca-journal/issues/2017/volume-4/enterprise-security-architecturea-top-down-approach`

- CIS website: `cissecurity.org`

12
Architecting for DevSecOps

As with everything in the enterprise IT domain, DevSecOps requires an architectural foundation. In this chapter, you will learn how to compose the reference architecture for DevSecOps practices and design the pipelines for DevSecOps. We will also discuss the best DevSecOps practices for the major public cloud providers; that is, AWS, Azure, and GCP. For that, we will elaborate on some of the leading tools in the market. In the last section, you will learn what steps the enterprise should take to implement DevSecOps.

After completing this chapter, you will be able to name the different components in a DevSecOps architecture and how to include these in a DevSecOps pipeline. You will have also learned how to secure containers and what the best practices are in various public clouds. Most importantly, you will be able to explain why including security in DevOps is crucial for enterprises.

In this chapter, we're going to cover the following main topics:

- Understanding the DevSecOps ecosystem
- Creating the reference architecture
- Composing the DevSecOps pipeline
- Applying DevSecOps to AWS, Azure, and GCP
- Planning deployment

Understanding the DevSecOps ecosystem

In the previous chapter, we discussed security principles and how this impacts the DevOps way of working. We concluded that security must be at the heart of every step in the development and deployment cycle, from the moment where code is pulled from a repository to the actual code commit and push to production. In this chapter, we will look at the foundation of **DevSecOps, DevOps that has security embedded**.

DevSecOps consists of three layers:

- **Culture**: This is not a technical layer, but it's often forgotten that DevOps is much more than just applying tools and creating CI/CD pipelines. Obviously, the same applies to DevSecOps. Within DevSecOps, every team member feels responsible for security and acts accordingly, taking ownership of it. This doesn't mean that security specialists have become obsolete, though. It's a good practice to have a security engineer or professional in the team, sometimes referred to as the security champion. This person must lead all processes in terms of applying security standards and policies, ensuring compliance.

- **Security by design**: Security is embedded at every layer of the system. This typically means that an enterprise has a defined architecture that covers every aspect of security and enforcing security postures onto systems: authentication, authorization, confidentiality, data integrity, privacy, accountability, and availability, including remediation and corrective actions when systems are under attack. Software developers do not need to think of security every time they design and build new applications or features – the posture is applied as soon as development starts. The security architecture, frameworks, rules, and standards are centrally managed.

- **Automation**: In DevSecOps, we want to automate as much as we can, and this includes security. The rationale for automating security is that we can prevent human error, and also have automated tollgates where code is scanned for possible vulnerabilities or non-compliant components such as unlicensed code. The security lead also takes responsibility for automating security, but does so with the team. Automation also implies automated audits and collection of evidence in case of attacks or breaches. Next, the automation process makes sure that security metrics are collected and sent back for feedback in the DevSecOps practice. For example, if, when you scan, a vulnerability in the code is discovered or a license has been breached, evidence will be collected and sent for feedback.

To manage these layers, DevSecOps relies on the following components:

- Harnessing repositories
- Application (code) security
- Cloud platform security
- Vulnerability assessments and testing

DevSecOps should not be mixed up with **security as a service** (**SECaaS**). SECaaS can be a component of the DevSecOps practice, but the concept of SECaaS is mainly about shifting security as a responsibility to a service provider. It's a sourcing model that allows enterprises to get cybersecurity delivered from a service provider on a subscription base. There are good reasons for implementing SECaaS, and one of them is that a provider is responsible for all security updates, based on the latest insights. Enterprises can define service-level agreements for incident response times and the timely application of security practices. As we mentioned previously, it can be integrated into DevSecOps, but SECaaS also means that an enterprise has to rely upon a third party for implementing and managing the security baseline.

In the next section, we will discuss the DevSecOps components and define the reference architecture.

Creating the reference architecture

Before we discuss the reference architecture of DevSecOps, we need to understand what the role of DevOps is and how security fits in. DevOps is about the software development life cycle. An important note that we have to make is the fact that developers increasingly use open source components. This makes sense since this provides great flexibility when developing new code.

Open source is community-driven, so developers can contribute to each other's code and speed up the process. Projects can and are shared in open Git and GitHub repositories, but also internally in enterprises. InnerSource type projects are a good example of this. InnerSource uses open source best practices for software development, within the boundaries of an organization. Typically, InnerSource projects make use of shielded, access restricted repositories in GitHub or alike.

Yet, there are some risks associated with open source that need to be addressed from a security perspective. Because of its open, community character – the strength of open source – there's an increased risk of introducing vulnerabilities to the code base. A second risk is license compliance. Licenses are not at the top of everyone's mind in open source, but be aware that even open source software and tools require licensing.

Let's look at the process first. The software development life cycle is a repetitive process. The developer pulls source code out of a repository and a build is triggered. After the code has been written, the code is packaged and enabled for deployment to the next stage in the promotion path; that is, test, acceptance, and eventually production. The whole process is facilitated through CI/CD pipelines and monitored. As we have concluded in the previous chapters, it's essential to test the code throughout the whole process. We also scan the code for security and compliance. This should be done at every single step in the process.

In fact, we need security from the start of the DevOps process. In practice, this means that we start scanning for security issues from the moment the code is pulled from the repositories. The repositories are indeed part of the software development life cycle too, so these must be protected from unauthorized access. This calls for **role-based access control (RBAC)** and **Identity and Access Management (IAM)** on repositories.

With that in mind, we can create the reference architecture for DevSecOps with the following components:

- Repository access with RBAC
- **Static Application Security Testing (SAST)**: This will detect errors in the source code
- **Software Composition Analysis (SCA)**: This will detect dependencies in code
- **Dynamic Application Security Testing (DAST)**: This will dynamically scan the code

These components are embedded in the DevSecOps pipeline, which we will discuss in the next section.

Composing the DevSecOps pipeline

Let's look at a common DevOps pipeline first. The basic pipeline is shown in the following diagram:

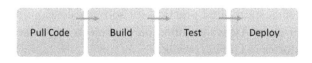

Figure 12.1 – DevOps pipeline

The basic steps in the pipeline are as follows:

- Pull code from the repository
- Build
- Test
- Deploy

In DevSecOps, we are embedding security into the pipeline, making security standards and policies an integrated part of it. Security is a layer that is applied to every step in the pipeline, but it does include several steps. This is shown in the following diagram:

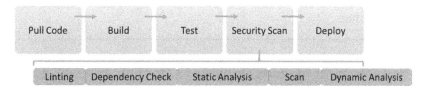

Figure 12.2 – DevSecOps pipeline

These steps are as follows:

1. **Dependency check**: First, any vulnerability that exposes the code to the risk of an exploit should be removed. This includes code that relies on other pieces of code to run. There are differences in code dependencies: developers can have controlled and uncontrolled dependencies. As a common practice, we don't want dependencies in our code. The risk is that if the code that the dependency relies on is breached or the function of that code is halted, the entire application might fail. Alternatively, malware that is injected in a certain piece of code, but has a dependency with other code, will infect the entire code stack. Therefore, the ground rule is to eliminate dependencies – this is one of the basic principles of zero trust, which we will discuss further in *Chapter 15, Implementing Zero Trust Architecture.*

   Package managers will check the code. Examples include `pipenv` for Python code and npm for Node.js. The commands that are used for the checks here are `pipenv check` and `npm audit`, respectively.

   > **Tip**
   > Check the `pipenv` website for scripts and tutorials on `https://pipenv.pypa.io/en/latest/`. Take a look at `https://docs.npmjs.com/cli/v6/commands/npm-audit` for npm code checks.

2. **Static analysis**: This checks for bad coding practices, such as bad configurations. There are open source tools for almost every coding language. Some examples of tools are as follows:

- ArchUnitNet and Puma Scan for C#

- Go vet for Go

- Checkstyle and **Open Web Application Security Project (OWASP)** dependency checks for Java

- Flow for JavaScript

- Parse for PHP

- Bandit for Python

> **Tip**
> This list is by no means exhaustive. On `https://github.com/analysis-tools-dev/static-analysis`, you will find a list of the current, most used tools.

3. **Scanning**: Developers will likely use containers and thus container images to build and package their applications. These images need to be scanned for vulnerabilities in used binaries and libraries. This scanning is done with base lists of known vulnerabilities; these lists are provided by institutes such as the **National Institute of Standards and Technology** (**NIST**), but also software providers in the form of **Common Vulnerability and Exposures** (**CVE**) notifications. As soon as a new CVE is reported, the lists are updated, and the scanning tools are automatically updated and triggered to redo the scan. Clair (`https://github.com/quay/clair`) is an open source tool that performs these scans, also for Docker images. Scanning involves **linting**, which we will explain in more detail when we talk about hardening containers in the next section.

4. **Dynamic analysis**: In the case of web applications, developers can run an automated web application scan to check for bad headers or missing tokens for **cross-site request forgery** (**CSRF** or **XSRF**). These tokens prevent exploits of unauthorized commands that come from a trusted user – this can also be a function on a different website. These automated dynamic scans can be integrated into the pipeline. The OWASP Zed Attack Proxy is a free web security tool (`https://owasp.org/www-project-zap/`).

Now, we have a security-embedded CI/CD pipeline that will automatically cover most commonly recognized vulnerabilities in code. There's one specific item that we didn't touch on in this section, though, and that's the use of containers and how we can secure these. We will study secured container builds in the next section.

Using secured containers in the pipeline

Most developers will use containers to wrap and deploy their code, typically Docker containers. There are some best practices when it comes to using and securing containers. To keep containers consistent and secured, they should be scanned regularly, even when the application has reached a steady state and updates are done less frequently or active development has stopped. If the application still runs with its underlying containers hosting the different application components, these containers must be scanned since there's always a possibility that a dependency is creating a new vulnerability.

Applications consisting of containers are defined by Dockerfiles. **Linting** – analyzing the code for errors or bad syntaxes used in the code – can be used to do **Static Code Analyzer (SCA)** of the Dockerfiles and make sure that these files remain secure. A popular linting tool to do this is **Haskell Dockerfile Linter (Hadolint)**. It's available as a Docker image and can easily be executed through the following command:

```
docker run --rm -i hadolint/hadolint
```

Hadolint will scan the code and if everything is all right, it will return an exit code of 0. When it discovers errors or bad practices, it will present a **Hadolint error** (DL) or **SellCheck error** (SC) key.

> Tip
> An overview of common errors is collected at https://github.com/hadolint/hadolint#rules.

Besides linting, Docker recommends some best practices for keeping containers secure. Docker already takes care of namespaces and network stacks to provide isolation so that containers can't obtain privileged access to other containers, unless specifically specified in the configuration. Next, there are some important things to consider:

- Docker uses the Docker daemon. This daemon requires root access, which implies security risks. First, only trusted users should be allowed to set controls for the daemon. Next, you will need to take action and limit the attack surface of the daemon by setting access rights to the Docker host and the guest containers, especially when containers can be provisioned through an API from a web server.

- The use of Docker Content Trust Signature Verification is strongly recommended. It's a feature that is available from the `dockerd` binary and allows you to set the Docker engine to only run signed images. For the signing itself, you can use Notary.

- Use hardened templates for Linux hosting systems such as AppArmor and SELinux.

If we follow up on all the recommendations of Docker, we will have tested, immutable images that can use to deploy containers on Kubernetes, for instance. Kubernetes will use the trusted image repository and takes care of provisioning, scaling, and load balancing the containers. One of the security features of Kubernetes is its support for rolling updates: if the image repository is updated with patches or enhancements, Kubernetes will deploy the new versions and destroy the previous ones. With this feature, developers will always be sure that only the latest, hardened versions of images are used.

Applying secrets management

Database credentials, API keys, certificates, and access tokens must be stored in a safe place at all times. The use of CI/CD and containers doesn't change that. It's strongly recommended to use a vault outside the repositories that the pipelines access for CI/CD. The best practices for secret management are as follows:

- Encryption at rest and in transit. AES-256 encryption keys are recommended.

- Secrets, such as keys, must never be stored in Git/GitHub repositories.

- It's advised that secrets are injected into the application via a secure string as an environment variable.

Hashicorp (Terraform) offers Vault as an open source solution for securely accessing secrets. The service allows us to easily rotate, manage, and retrieve database credentials, API keys, and other secrets throughout their life cycles.

A more robust solution is provided by CyberArk. CyberArk Conjur is a platform-independent secrets management solution, specifically architected for securing containers and microservices. The solution is platform-agnostic, meaning that it can be deployed to any cloud or on-premises system.

Both tools integrate with native environments for key management in, for example, Azure and AWS, which use Azure Key Vault and AWS Secrets Manager, respectively.

Applying DevSecOps to AWS, Azure, and GCP

In the previous sections, we discussed the DevSecOps principles and how the pipeline is built with embedded security. In this section, we will look at the best practices of applying DevSecOps to the major public cloud platforms, that is, AWS, Azure, and **Google Cloud Platform** (**GCP**).

Working with DevSecOps in AWS CodePipeline

Before we start exploring DevSecOps in AWS, we need to understand that deployments in AWS should be based on the principles of the **Cloud Adoption Framework** (**CAF**). That framework covers specific security tasks and responsibilities, grouped into the four categories or principles for enterprise security that we discussed in *Chapter 11, Understanding Security in DevOps*:

- Prevention
- Detection
- Correction
- Direction

> **Note**
> AWS refers to these principles with different terminology for correction and direction. In CAF, these are subsequently called detective and responsive.

AWS offers native solutions to provide controls for managing security postures in CI/CD pipelines: Amazon CloudWatch Alarms, AWS CloudTrail, Amazon CloudWatch Events, AWS Lambda, and AWS Config. The following diagram shows the CI/CD pipeline for DevSecOps using these solutions:

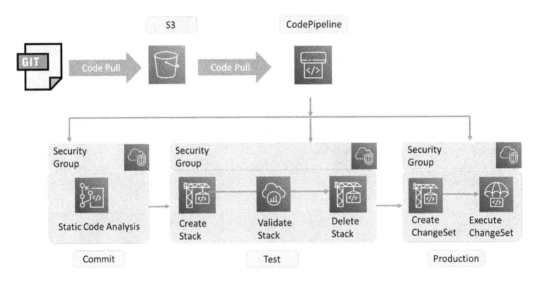

Figure 12.3 – Using CodePipeline and security groups in AWS

AWS CodePipeline is used to orchestrate the different steps in the pipeline. An important artifact is the security groups: these are the *bins* where the security posture of all the components that are developed and deployed in the pipeline is defined. It contains the templates, guardrails, and policies that have to be applied to these components. We can define three stages in the pipeline:

1. **Source or commit**: Static code analysis is performed on the code that is pulled from an S3 bucket. In the case of security group breaches, the build will be stopped.

2. **Test**: In this stage, CloudFormation is used to create a stack that contains a **Virtual Private Cloud (VPC)** in AWS to run the tests. Next, AWS Lambda is used to run the code in the stack and validate the build. AWS calls this stack validation: Lambda functions will validate the stack against the security groups. If a breach is detected, a Lambda function will delete the stack and send out an error message. This is to prevent the stack and the code from entering the next stage.

3. **Production**: After a successful stack validation, a Lambda function is triggered to prepare the stack for production using CloudFormation templates. This *change set* – translating the test stack into production with production templates – is then executed.

> **Tip**
> AWS provides samples for CloudFormation templates and pipelines at
> `https://github.com/awslabs/automating-governance-`
> `sample/tree/master/DevSecOps-Blog-Code`.

Examples of items that are checked against security groups can be validating user access and permissions, access controls to the S3 buckets, and the policies to create instances using, for example, EC2 compute resources. CloudWatch and CloudTrail are used to monitor the components, the access levels, and their usage, and will collect logs from events that are triggered during the execution of the various steps in the pipeline.

Working with DevSecOps using GitHub and Azure services

Microsoft Azure uses a different approach to DevSecOps: it leverages the scan possibilities of GitHub and the features of **Azure Kubernetes Services** (**AKS**), next to Azure Pipelines, which is integrated into Azure DevOps and Azure Security Center for storing the security postures. The following diagram shows a high-level architecture for a security embedded CI/CD pipeline using GitHub and Azure services:

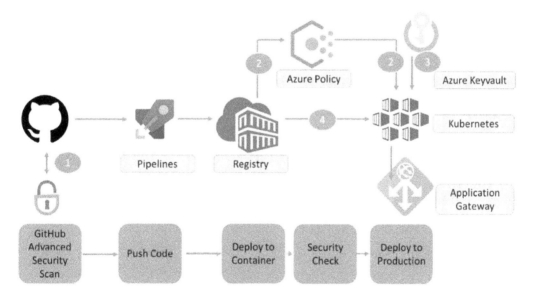

Figure 12.4 – DevSecOps with GitHub and Azure services

The numbers in the preceding diagram represent the order in which steps are taken. As soon as the containers are pushed to **Azure Container Registry (ACR)**, they are scanned against the policies that are stored in Azure Policies. Next, the appropriate security keys are fetched to authenticate the containers to **Azure Kubernetes Service (AKS)**. Only when all the checks have passed will the code be pushed to the application gateway.

Let's look at this in a bit more detail:

1. **Source**: The solution starts with code analysis in GitHub, which involves using CodeQL and Dependabot to detect vulnerabilities in the source code and dependencies, respectively.

2. **Test**: Once the code has been validated, it's packaged in a Docker container and deployed to a test environment using Azure Dev Spaces. This orchestration is done through Azure Pipelines. Azure Dev Spaces will build an isolated test environment using AKS. This is comparable to how CloudFormation in AWS builds stacks.

3. **Scan**: Containers are stored in the ACR, where they are scanned against the security posture. For this, Azure uses Azure Security Center, which is a huge library that holds all security policies for environments that are enrolled in Azure.

4. **Production**: Scanned containers are pushed to a Kubernetes cluster using AKS. Azure Policies are used to validate the compliance of provisioned clusters and containers.

Just like AWS, Azure uses several different solutions to provide an end-to-end solution that embeds security rules, policies, and postures throughout the whole CI/CD process. However, all these solutions start with a repository where these security guardrails and guidelines are stored and managed: security groups managed through AWS Security Hub or, in Azure, the Azure Security Center.

Working with DevSecOps in Google Cloud using Anthos and JFrog

GCP offers an interesting best practice solution for implementing DevSecOps pipelines using Anthos and JFrog. With this, it doesn't only provide a cloud-native pipeline, but also a solution to develop and deploy for hybrid environments, using GCP and on-premises systems.

This is interesting for enterprises since a lot of enterprises will not move their IT systems completely to public clouds. Most enterprises are expected to move more and more systems to the cloud, but some of their systems will remain on private stacks. CI/CD pipelines that cater for both cloud and on-premises solutions are favorable and with Kubernetes, they are relatively easy to set up.

The architecture is shown in the following diagram:

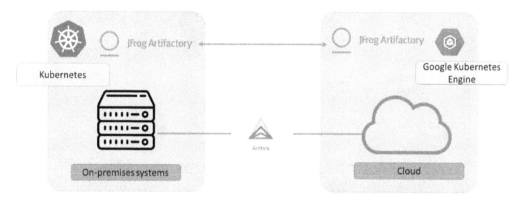

Figure 12.5 – High-level architecture of using JFrog Artifactory and Google Anthos

GCP advocates the use of JFrog Artifcatory and JFrog Xray:

- **JFrog Artifactory** takes care of storing artifacts that are used when building applications. In this chapter, we saw that a pipeline starts by pulling code from source repositories. Developers need to be able to rely on the tooling that stores and orders artifacts – code building blocks – comprehensively and safely so that software delivery to the pipelines can be automated.

- **JFrog XRay** scans the artifacts – the code building blocks – through Artifactory against known vulnerabilities and license compliance. XRay advocates the shift-left mentality by already scanning the source artifacts.

The solution is shown in the following diagram:

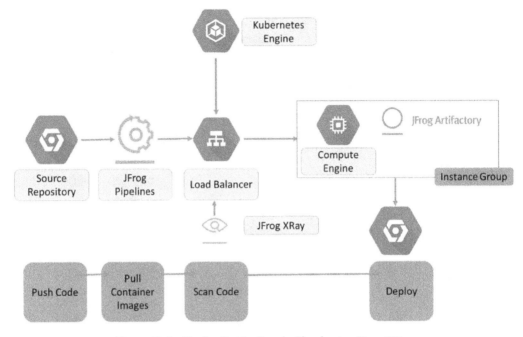

Figure 12.6 – DevSecOps in Google Cloud using JFrog XRay

In this solution, JFrog XRay is the security solution that is embedded in the pipeline. Builds are then pushed to production, using Kubernetes in GCP, and on Anthos. Anthos, however, ensures a consistent layer for deploying and managing Kubernetes clusters across the native cloud with **Google Kubernetes Engine** (**GKE**) and on-premises. This solution is not only feasible with GCP, but it can be used on top of VMWare stacks on-premises, as well as on AWS.

Planning for deployment

So far, we've discussed the reference architecture for DevSecOps pipelines and the best practices for AWS, Azure, and GCP. If we have the architecture, the next step would be planning to deploy DevSecOps and the pipelines in our enterprise. That's the topic of this final section.

There are three major steps that enterprises will need to follow to implement DevSecOps:

1. **Assess the enterprise security**: Enterprises will likely already have adopted security policies and taken measures to protect their systems. They will also need to adhere to security standards and frameworks, because of governmental or industry regulations. Security specialists will have conducted risk assessments and analyzed possible threats. These specialists understand and manage the security controls. This is, by default, the starting point of merging security into the DevOps practice. A very strong recommendation is that DevOps teams should not start without including security policies and standards for developing and deploying new code, not even in pilot projects or Proof of Concepts. Security must be a top priority from day 1.

2. **Embed security into DevOps**: Security policies and standards are integrated into the development process. The DevOps workflows are matched against the security guidelines and guardrails. This includes vulnerability testing and code scanning, which we discussed extensively in this chapter. Without processes and tools in place, DevOps teams can't start developing new code. The risk of increasing the attack surface of systems and, ultimately, causing immense damage to the enterprise is too big. Companies, both big and small, are under the constant threat of hackers and security threats. That brings us to step three.

3. **Train, train, train**: DevOps and DevSecOps aren't only about technology – it's a way of working and even thinking. Maybe even better formulated: it's a culture, and people need to be trained in adopting that culture. That training is not a one-off. Staff, developers, and operators need to be trained constantly and consistently. Developers, operators, and security engineers need to be fully committed to applying the security controls throughout their work, and that implies that they always need to be aware of the risks an enterprise is facing in terms of security breaches and hacks.

Of course, proper tooling is essential. Enterprises are recommended to include the following tools as a minimum:

- **Testing**: This is the crucial element in DevSecOps. The market provides a massive number of tools for performing tests. Examples are Chef Inspec, Haikiri, and Infer.

- **Alerting**: When security threats are detected, alerts need to be raised and sent out. Elastalert is an example of an alerting tool.

- **Automated remediation**: Tools such as StackStorm can help in providing remediation as soon as security issues are detected.

- **Visualization**: Developers and operators need to be able to see what's going on in systems. Grafana and Kibana are popular tools that help in visualizing and sharing security information.

This list is by no means intended to be exhaustive. The tools mentioned are third-party tools that integrate well with DevOps tooling and native tooling in AWS, Azure, and Google Cloud. Of course, the public cloud platforms themselves offer extensive security tooling. Examples are Sentinel and Azure Security Center in Azure, Security Hub in AWS, and the Security Command Center in GCP.

The benefits of DevSecOps should be clear after reading this chapter, but we will summarize this with a conclusion: with DevSecOps enterprises, we can achieve better collaboration between developers, operators, and security engineers and with that, ensure that security threats and vulnerabilities are detected at an early stage of development so that risks for the enterprise are minimized.

We will elaborate on implementing security in DevOps in *Chapter 14, Integrating DevSecOps with DevOps*, where we will also discuss DevSecOps governance. But first, we will learn how to work with and integrate industry security standards in DevOps in the next chapter.

Summary

In this chapter, we studied the different components of DevSecOps. We learned that DevSecOps is not only about tooling and automation, but also very much about culture: DevOps teams have to collaborate with the security specialists in the enterprise and together, they must be fully committed to embracing and embedding security guidelines into developing and deploying new code. Tools can certainly help in achieving maximum security in DevOps. A larger part of this chapter was about architecting the DevSecOps practice.

Then, we discussed the best practices for DevSecOps in the major public cloud providers; that is, AWS, Azure, and Google Cloud. These practices typically include the use of Docker containers and Kubernetes as container orchestration platforms. We also learned how to scan code and secure the containers before deploying them to a production platform. Important activities include static code analysis and dynamic scanning.

In the last section of this chapter, we discussed the steps an enterprise must take to implement the DevSecOps practice and provided some recommendations for the necessary tools.

Enterprises must typically adhere to governmental and industry security standards and frameworks. The next chapter is all about working with these standards in DevSecOps.

Questions

1. What is the function of **software composition analysis (SCA)**?

2. What technique is used to keep containers secure?

3. What is the native tool in AWS that's used to create stacks?

4. The AWS, Azure, and GCP public cloud providers offer their own Kubernetes services to run containers. Name their respective services.

Further reading

- Blog on using AWS CodePipeline in DevSecOps: `https://aws.amazon.com/blogs/devops/implementing-devsecops-using-aws-codepipeline/#:~:text=%20Implementing%20DevSecOps%20Using%20AWS%20CodePipeline%20%201,%206%20Create%20change%20set%3A.%20%20More%20`

- Documentation on applying DevSecOps practices in Azure: `https://azure.microsoft.com/en-us/solutions/devsecops/`

- Documentation on DevSecOps CI/CD using GCP, Anthos, and JFrog: `https://cloud.google.com/architecture/partners/a-hybrid-cloud-native-devsecops-pipeline-with-jfrog-artifactory-and-gke-on-prem#best_practices`

- Documentation on security in Docker: `https://docs.docker.com/engine/security/trust/`

13
Working with DevSecOps Using Industry Security Frameworks

An important artifact in security – and DevSecOps – is security frameworks. There are generic frameworks, such as **Center for Internet Security** (**CIS**), but typically, industries must comply with and report about compliancy according to specific industry security standards. These have an impact on the way security is handled within enterprises and therefore in the implementation of DevSecOps.

This chapter will explain the functionality and impact of frameworks and how to incorporate them into DevSecOps. This chapter includes a separate paragraph on the use and value of the MITRE ATT&CK framework since it is becoming more well-known and more widely accepted as a base framework.

After completing this chapter, you will have a good understanding of the most used security frameworks and how the controls of these frameworks can be applied to DevOps.

In this chapter, we're going to cover the following main topics:

- Understanding industry security frameworks

- Working with the MITRE ATT&CK framework

- Applying frameworks to DevSecOps

- Creating compliance reports and guiding audits

Understanding industry security frameworks

IT has become more complex over the years. The same applies to IT security. There's a correlation between the two. Enterprise IT environments are no longer monolithic systems that sit in the basement of a company that's functioning as the enterprise's data center. Today, IT environments share different components and have connections to the outside world through internet connections. With that notion, systems are, *by default*, accessible through the internet. Yet, only authorized users should be able to access these systems. Hence, we need some strong defenses to protect systems from security breaches.

The level of required security will differ per industry. First of all, financial institutions will want to make sure that bank accounts can't be compromised and that money is not being illegally transferred. Healthcare institutions need to protect their patients' personal and health data. Manufacturers want to protect their intellectual property and patents. Above all, there are several overarching principles in terms of security, protecting data, identities, and hardening systems from outside attacks. It's almost impossible to keep track of all this, and that's where security frameworks come in: they provide guidance for implementing the right set of security policies in an enterprise.

Before we learn how security frameworks impact CI/CD and DevOps, we will need to understand what these frameworks are. In short, a framework is a set of policies and documented guidelines on implementing and managing these policies. The policies themselves are focusing on identifying risks, mitigating risks, and reducing the attack surface of systems and procedures in case vulnerabilities are detected. This is a generic approach, but industry frameworks tune this approach to specific needs in an industry.

Generic IT security frameworks include ISO IEC 27001/ISO 2700212, the **National Institute of Standards and Technology (NIST)** Cybersecurity Framework, **Center for Internet Security (CIS)**, and **Control Objectives for Information and Related Technologies (COBIT)**. Let's explore these in a bit more detail first:

*   **ISO IEC 27001/ISO 2700212/27017**: ISO 27001 is setting international standards for system security controls. The emphasis is on controls that detect threats that will have a severe impact on the availability and integrity of systems. ISO 27002 sets additional standards to manage the controls themselves, such as user access management and maintaining asset inventories. ISO 27017 specifically targets the cloud. It deals with shared responsibilities in **Platform-as-a-Service (PaaS)** and **Software-as-a-Service (SaaS)** environments, secures deployments and removes cloud systems, and monitors cloud services, as a few examples.

*   **NIST**: The NIST Cybersecurity Framework doesn't specify controls, but it does provide five functions to enhance security: identify, protect, detect, respond, and recover. These functions allow organizations to set controls to manage data breach risks. Controls must include access control, measures to protect data, and also the awareness of staff. Respond calls for controls must describe how organizations should react to threats and attacks, including mitigation and communication guidelines. Recovery is the last resort: organizations need to have a clear strategy regarding how to recover from attacks, such as system and data recovery. The five NIST domains are shown in the following diagram:

Figure 13.1 – The NIST cybersecurity framework

- **CIS**: CIS offers extensive frameworks with specific controls for platforms, operating systems, databases, and containers. Some of the CIS frameworks are embedded in platforms such as CIS for Azure and AWS. In these scenarios, the CIS benchmarks can be accessed from Azure Security Center and AWS Security Hub. The CIS benchmarks make sure that used components are *hardened*. The big difference compared to NIST is that NIST focuses on guidelines to assess risks, whereas CIS provides long lists of security controls and best practices.

- **COBIT**: COBIT was launched by **Information Systems Audit and Control Association (ISACA)**, the international organization of IT security and audits. Originally, COBIT was about identifying and mitigating technical risks in IT systems, but with the recent release of the framework – COBIT 5 – it also covers business risks that are related to IT. COBIT is complicated to implement and manage since it covers the entire enterprise, including all IT management processes such as incident, problem, configuration, and change management.

These are control frameworks. All of these frameworks may have *editions* that cover specific industry requirements, but typically, industries have to adhere to their *own* standards, as well as be fully compliant. This is important when industries are audited. In the last section of this chapter, we will discuss auditing in more detail.

The major industry frameworks – actually, these are regulatory attestations – are **Health Insurance Portability and Accountability Act (HIPAA)** for healthcare, **Federal Risk and Authorization Management Program (FedRAMP)** for governmental organizations in the US, **General Data Protection Regulation (GDPR)** in the European Union, and **Payment Card Industry Data Security Standard (PCI-DSS)** for financial institutions.

All of these are global or at least regionally implemented, but there may also be specific national security regulations a company needs to apply. An example is the **New York Department of Financial Services (NYDFS)** Cybersecurity Regulation. This framework was released in 2017, placing security regulations on all financial institutions in the US. However, the rules in this framework are aligned with NIST and apply ISO 27001 standards. Yet, NYDSF does have some rules that supersede these generic frameworks. Under NYDSF, data encryption and enhanced multi-factor authentication are mandatory security controls for all inbound connections.

There's one framework that we haven't discuss yet and that's MITRE ATT&CK. MITRE ATT&CK is not a real framework, such as the ones that we discussed in this section. It's a knowledge base that covers tactics on how systems might be attacked and breached. It can, however, be used as input to define risk strategies and threat models to protect systems. In the next section, we will learn how to use MITRE ATT&CK.

Working with the MITRE ATT&CK framework

Maybe it's not a completely fair statement, but we will post it here regardless: MITRE ATT&CK lets you think from the attacker's perspective when it comes to security. The strength of this framework is that anyone can contribute to it. It doesn't really describe the actual vulnerabilities in systems, but more the techniques attackers could use to exploit these vulnerabilities. MITRE ATT&CK uses a matrix with 14 attack tactics. Next, it divides these tactics across major platforms or technologies, including cloud and containers. In the cloud, there's a subdivision for Azure, AWS, and GCP.

> Tip
>
> The full MITRE ATT&CK framework can be found at `https://attack.mitre.org/`. However, it is recommended to follow MITRE on Twitter as well at `@MITREattack`. The matrix is open source, so a lively community is contributing to the tactics and techniques that are collected in the framework. MITRE invites people to join the community and actively contribute to their findings.

In this section, we will briefly go over the 14 tactics and then specifically address the ones for containers, since these are widely used in **Continuous Integration/Continuous Deployment (CI/CD)**:

- **Reconnaissance**: These techniques gather as much information as possible to prepare attacks. This includes scanning systems, but also social engineering, wherein the staff of organizations are used to get information.

- **Resource development**: These techniques involve creating, purchasing, or stealing resources that hackers can use to execute an attack.

- **Initial access**: This is the first attempt at gaining access to systems. This includes abusing valid (service) accounts and phishing.

- **Execution**: This technique is about running malicious code.

- **Persistence**: This technique gains access using backdoors. It uses containers to inject malicious code, boot, and log into initialization scripts.

- **Privilege escalation**: This technique involves using exploits to leverage privileges on systems to eventually gain more control. When it comes to using containers, this is a commonly used tactic. Containers should always be hardened to prevent them from gaining extra privileges.

- **Defense evasion**: This involves tactics wherein code is used to go around intrusion detection, logging, and other prevention measures. In cloud environments, this tactic is used to manipulate cloud (coded) firewalls by, for example, entering through an unused environment in a different region or unprotected sandbox environments.

- **Credential access**: Typically, this involves brute-force attacks to get usernames and passwords.

- **Discovery**: This tactic is used to find user data, devices, applications, data, and services to gain as much information about the available systems as possible.

- **Lateral movement**: This tactic is used to move systems and data from one host to another, sometimes in a different environment that's not under the control of the enterprise. Pass the hash and remote admin access are commonly used techniques.

- **Collection**: This tactic is used to collect data from keyboard strokes or screen captures, for instance. In the cloud, collecting API keys to access storage and key vaults are popular techniques.

- **Command and control**: When using this technique, hackers try to communicate with systems in an attempt to gain control of them.

- **Exfiltration**: This tactic involves gaining control of data and data streams by, for instance, sending data to different storage environments and encrypting the data.

- **Impact**: This is a broad category and includes denial of service techniques and resource hijacking.

MITRE ATT&CK is not a magic wand: it doesn't solve all security problems. It should be considered another source you can use to start protecting IT environments in a better way by providing different insights. It shows potential attack patterns and paths that security engineers can include in their security policies. MITRE ATT&CK provides insights from specific platforms and technologies, which makes it rather unique. Common attack tactics may have different paths and patterns in various platforms, and that's where MITRE ATT&CK is a good guardrail.

Note

In DevOps, scanning code is crucial. Static and dynamic scanning must be the default in CI/CD. Scanning is typically done against baselines. One baseline that is commonly used in DevOps is **OWASP**, the **Open Web Application Security Project**. OWASP is open source and yearly lists the top 10 vulnerabilities in applications. We will discuss OWASP in more detail in *Chapter 14, Integrating DevSecOps with DevOps*.

In the next section, we will show you how the matrix can be used to better protect containers.

Using MITRE ATT&CK tactics for containers

So, how does MITRE ATT&CK work in practice? Let's use the matrix for containers, since they are frequently used in CI/CD pipelines, as an example. First, we must go to the specific matrix on `https://attack.mitre.org/matrices/enterprise/containers/`. You will recognize some of the 14 tactics that we discussed in the previous section. Not all of them are applicable; for containers, eight tactics have proven to be relevant:

- Initial access
- Execution
- Persistence
- Privilege escalation
- Defense evasion
- Credential access
- Discovery
- Impact

Next, we can look at one of these tactics. We will use **Execution** as an example, as shown in the following screenshot:

Figure 13.2 – Tab in MITRE ATT&CK for containers

The top two in Execution are as follows:

- **Container Administration Command**
- **Deploy Container**

If we click on the first one, **Container Administration Command**, the matrix provides us with information on how a specific vulnerability in container administration has been exploited. The vulnerability itself may occur while managing the containers using, for instance, a Docker daemon or the Kubernetes API. These may allow remote access and management to be used while the containers are launched. MITRE gives two examples of techniques that have been used for this. The first technique is Hildegard, which makes it possible to execute commands on running containers using the kubelet API run command. Kinsing, the second technique that MITRE mentions, exploits an Ubuntu entry point to run shell scripts to take over container administration processes.

Next, the matrix provides mitigations. In the given examples, the mitigating actions are using read-only containers and limiting communications with the container service, daemon, or Kubernetes using remote access via SSH.

You will find that Kinsing also appears under the **Deploy Container** tab, next to the Doki exploit, which is a piece of malware that was discovered in the spring of 2020 and targets Docker containers.

The matrix will guide you through various exploits and help you mitigate them.

With that, we've discussed the contents of the various frameworks. In the next section, we will learn how to use them in DevSecOps and how compliance reports are created to show that frameworks have been applied.

Applying frameworks to DevSecOps

In this section, we will learn how to include the controls of frameworks in DevOps and embed them as DevSecOps. Good news: it's not as hard as it may sound. The following diagram shows this process:

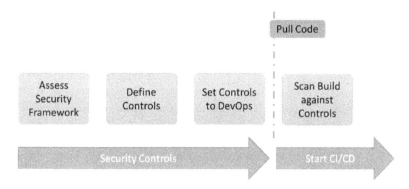

Figure 13.3 – Process of applying controls from security frameworks to DevOps

In general, we start by assessing the framework that the enterprise needs to apply to their IT environments. From that assessment, the different controls are derived and set to the development and deployment cycles of applications and infrastructure. As soon as code is pulled from the repositories, scanning starts against these controls.

We are using the CIS benchmark as an example here since CIS is the most used framework for setting security controls. Applying controls starts with the realization that in DevOps, the IT environments are highly dynamic by default. Everything, including the infrastructure, is turned into code, so applications will run in containers or in serverless mode. This calls for some specific controls.

Some generic controls must be applied. These include the following:

- **Vulnerability management**: This must be implemented as a control before code is pushed to production, but with the principles of shift-left in mind, vulnerability scanning should already start from the moment that code is pulled from repositories.

- **Access**: With this control, you can limit and manage the privileges of all resources, including containers.

- **Logging**: This includes the logs while building and testing code. Sometimes, only logs are collected in production environments, but that is not sufficient if you want to be in control of the DevOps cycle.

These are generic. CIS has developed a specific framework for securing containers, as shown in the following figure:

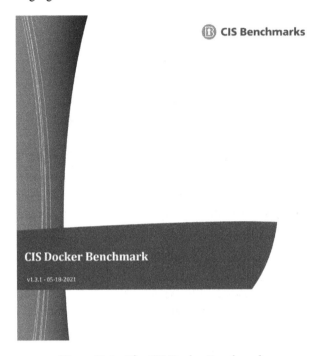

Figure 13.4 – The CIS Docker Benchmark

This benchmark, as CIS calls it, contains controls for the following:

- Linux host configuration; for example, ensuring a separate partition for containers and ensuring that only trusted users can control the Docker daemon.

- Docker daemon configuration; for example, running the daemon as a non-root user if possible and ensuring that containers are restricted in gaining new privileges.

- Container images and build file configuration; for example, ensuring that a container only uses trusted base images.

- Container runtime configuration; for example, Linux kernel capabilities are restricted in containers and ensuring that privileged ports are not mapped in containers.

CIS has some generic recommendations too, such as making sure that containers are hardened and that the Docker version is up to date.

> **Tip**
>
> All CIS benchmarks can be downloaded for free at `https://www.`
> `cisecurity.org/`.

The benchmark not only tells you *what* controls should be in place but also gives recommendations on *how* to implement these, together with the rationale of *why* the controls should be there. Well, look at the example where CIS recommends having a separate partition for containers – control 1.1.1 in CIS v1.3.1 (2021) for Docker.

This starts with *profile applicability*. In control 1.1.1, this is set to Level 1-Linux host. This means that the settings are only applicable to the Linux host, providing a clear security benefit without hampering the intended functionality of a component – in this case, the Linux host.

Next, the control itself is described, as well as the rationale behind it. In this example, it describes how Docker uses `/var/lib/docker` as the default to store all its components. The directory is shared with the Linux host, which means it can easily be filled up completely, making both Docker and the host unusable. Hence, a separate partition is recommended. Lastly, CIS provides a *manual* on how to do this by creating a separate partition for the `/var/lib/docker` mount point.

Do you have to follow through on all these recommendations? No. CIS has made a clear distinction between critical and important controls. Obviously, the critical ones should be implemented in all cases, but you will need to assess every control, regardless of whether it makes sense to implement them in your DevOps practice. The golden rule here is that if you implement a control, you need to adhere to it consistently and report on it for compliance. Enterprises will be audited on rules and policies they have implemented. In the last section of this chapter, we will discuss reports and audits.

Creating compliance reports and guiding audits

DevOps is taking a huge flight in enterprises. Embedding security in DevOps is a logical next step. But how can enterprises be sure that their DevOps and DevSecOps are compliant with the frameworks that we've discussed throughout this chapter? The answer to that question is: by audit. IT systems are regularly audited, and so should DevOps practices. Having said that, auditing DevOps is still *unchartered territory*, although major accounting firms such as KPMG and Deloitte have issued white papers on the subject.

DevOps audits should include at least the following topics:

- **Evaluating the DevSecOps strategy**: Is the strategy clear? How is governance arranged? A DevOps strategy can be set per business unit or enterprise-wide. Both are fine, so long as the strategy is followed through consistently. The goals should be clear and adopted by every team. The same applies to the way of working across all disciplines in the team. Processes such as testing procedures and acceptance criteria must be transparent and adhered to without exceptions.

- **Assessing the level of DevSecOps training**: Training is not just simply creating a presentation with a one-slider on **Scaled Agile Framework (SAFe)** and showing the DevOps cycles. DevOps is very much about culture, but sometimes, organizations simply get overwhelmed by a sudden new way of working. For example, implementing DevOps also means creating teams with the right skills. This needs to be organized and goes further than just releasing the **Spotify model** in an organization. Staff don't organize themselves in guilds and squads just by telling them they must. An enterprise will need to train its staff in DevOps and make sure the teams have the right skill sets. Training also includes managing organizations.

> **Note**
>
> The Spotify model has become immensely popular in organizations as an approach to scaling the agile way of working, which DevOps is part of. The Spotify model, named after the agile way of working that was implemented by the audio-streaming service, advocates autonomy in teams, organized into squads. Each squad is allowed to choose its own toolsets and agile framework, such as Scrum or Kanban.

- **Reviewing the DevSecOps toolchain**: Is there an architecture that specifies the DevOps tools, and is it coherent? Does it serve the strategy and is it aligned with the IT strategy of the enterprise? For example, if the enterprise has an open source strategy, then the tools must adhere to that. Lastly, just like any tool used in the enterprise, it needs to be put under architecture change control.

- **Reviewing the DevSecOps processes**: DevOps doesn't mean that processes are not valid anymore. Enterprises will still need to have the basic IT processes in place, such as incident management, problem management, and change management. These processes must be documented, including their escalation levels. Also, clear descriptions of the roles in these processes must be provided and followed through when implementing DevSecOps. Security management takes a special position here in that it must describe how security policies are defined, how they are implemented and managed in the enterprise, and how they are embedded in the DevOps process.

With that, we've studied the basic principles of security in DevOps and the industry frameworks for security. Now, we need to merge – or rather integrate – security into our DevOps practice. That's the topic of *Chapter 14, Integrating DevSecOps with DevOps*.

Summary

In this chapter, we discussed various security frameworks. These frameworks are guidelines for setting security controls for the IT environments of the enterprise. These controls apply to systems and applications, and also to the DevOps practice. From the moment developers pull code from a repository and start the build, up until deployment and production, IT environments, including CI/CD pipelines, need to adhere to security controls. There are a lot of different frameworks. Some of them are generically and broadly accepted by enterprises, such as NIST, CIS, and COBIT.

We also discussed the MITRE ATT&CK framework, which takes a different angle by comparing itself to other security control frameworks. MITRE ATT&CK lists tactics and techniques that hackers may use or have used to exploit vulnerabilities. Just like CIS, MITRE ATT&CK lists specifics for various platforms and technologies, including containers that are commonly used in CI/CD.

In the last section, we looked at auditing DevSecOps. It's recommended to review topics such as the consistent usage of tools, processes, and the skills of the DevOps teams.

In the next chapter, we will integrate the security practice into DevOps and learn how enterprises can adopt a true DevSecOps strategy.

Questions

1. What ISO standard is specifically for the cloud?

2. What two techniques does MITRE ATT&CK mention for containers under the execution tactic?

3. True or false: CIS doesn't mention the versioning of Docker as a control.

Further reading

- KPMG, January 2020: `https://advisory.kpmg.us/articles/2020/role-of-internal-audit-devops.html`

- Discover the Spotify model, a blogpost by Mark Cruth on Atlassian: `https://www.atlassian.com/agile/agile-at-scale/spotify#:~:text=It%20is%20now%20known%20as%20the%20Spotify%20model.,by%20focusing%20on%20autonomy%2C%20communication%2C%20accountability%2C%20and%20quality`

- Website of CIS: `https://www.cisecurity.org/`

- Website of ISACA, where the COBIT 5 framework can be found: `https://www.isaca.org/`

- Website of NIST: `https://www.nist.gov/`

- Website of ISO: `https://www.iso.org/standards.html`

14
Integrating DevSecOps with DevOps

The title of this chapter may sound a bit odd, but DevSecOps and DevOps aren't separate things. It should be one way of working: security should be integrated with the DevOps practice, instead of security principles being added on top of DevOps. This means that architects have to define one overarching governance model, integrate threat modeling into DevOps, and aim for an integrated toolset. Lastly, integrated monitoring needs to cover every aspect of the DevSecOps cycle. We will learn that integrated monitoring comes close to something that we discussed earlier in this book: AIOps. In this chapter, we will pull everything together.

After completing this chapter, you will have learned how to implement governance, understand threat modeling, and understand the importance of it in the secure **software development life cycle** (**SDLC**). You will have also learned how security is embedded into continuous integration and how this is monitored, as well as about some of the major tools in this domain.

In this chapter, we're going to cover the following main topics:

- Defining governance in DevSecOps
- Understanding and working with threat modeling
- Integrating tools and automation
- Implementing monitoring

Defining governance in DevSecOps

So far, we have drafted a DevSecOps architecture, identified processes, and then aligned these with the business goals of the enterprise. The next step is to manage all this, and that's the subject of governance. DevSecOps is not just a PowerPoint presentation and a Visio diagram showing the CI/CD pipelines. An enterprise needs skilled staff to work with it and a governance model that describes the secured digital operating model. In this section, we will discuss this by using the IT4IT framework by The Open Group as a best practice.

In *Chapter 6, Defining Operations in Architecture*, we introduced value streams for products and described how IT creates value. The model can be seen in the following diagram:

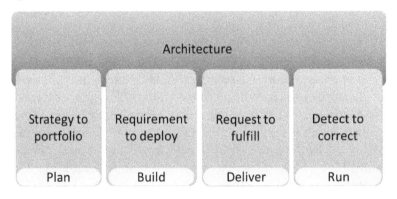

Figure 14.1 – IT4IT value streams

In IT4IT, **Governance, Risk, and Compliance (GRC)** is a supporting activity for the four value streams. This means that GRC is fully embedded in every value stream. What does GRC do?

Simply put, GRC is about achieving objectives while managing uncertainties by applying agreed and accepted industry business policies. Every business needs to adhere to certain policies. These can be international regulations for trading, national laws, and also industry-specific standards. This includes security and data privacy regulations. Implementing GRC is not a one time-effort: enterprises will need to adjust regularly. That's where the GRC capability model comes in. This model is shown in the following diagram:

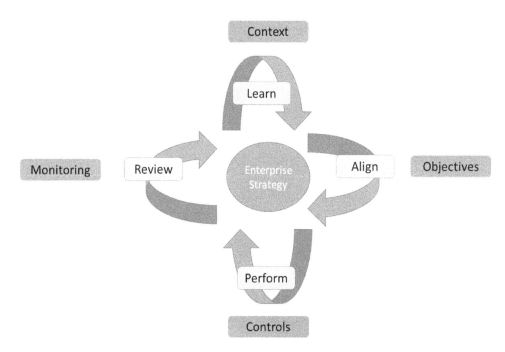

Figure 14.2 – The model for Governance, Risk, and Compliance

This model consists of four elements:

- **Learn**: The objectives of the enterprise must be clearly defined, but these are set in the context wherein the enterprise is operating. For example, the objective of a hospital is making people better; the context is much broader and includes, for instance, relationships with pharmaceutical companies and health care insurance companies. The objectives and the context define the strategy of a company.

- **Align**: The objectives of the enterprise and potential threats that endanger the strategy of the enterprise are assessed. This leads to requirements and countermeasures so that the enterprise can achieve its objectives and grab new opportunities while mitigating the threats.

- **Perform**: This is the stage where undesired events are detected and actioned.

- **Review**: The enterprise needs to constantly review if threats have been identified in time, rated appropriately, and that mitigating actions have been successful. This feeds back into learning, since the context might change and the strategy will need to be adapted.

Implementing and managing these capabilities requires governance. An enterprise will have to assign people to control these capabilities and the level of adoption of these capabilities at different levels in the enterprise. This includes integrating GRC capabilities into DevSecOps. How can we arrange a best practice governance model?

We will need to look at the different levels in the enterprise, and that's what IT4IT does. The following diagram shows the high-level principles of the model:

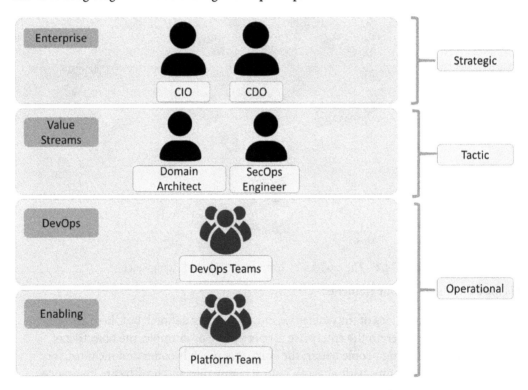

Figure 14.3 – Model for different security governance levels in the enterprise

Let's look at these levels in more detail:

- **Enterprise level**: The key role at enterprise level is the enterprise architect, who is responsible for the architecture in the enterprise. At this level, the enterprise architect will work together with the **Chief Information Officer** (**CIO**). In modern enterprises, we see the role of the CIO changing and other roles being added. The **Chief Digital Officer** (**CDO**) and the Chief Data or Chief Privacy Officer appear more and more on the organizational charts. The CDO is an important position for implementing the digital transformation strategy, including the adoption of DevSecOps. Be aware that DevSecOps is never a strategic goal in itself: it's a methodology that guides digital transformation.

- **Value streams**: The enterprise level is the strategic level; the value streams are the tactical level. The critical roles at this level are the domain architect and the SecOps engineer. They need to oversee the implementation of overarching DevOps architectures and security in the various DevOps teams.

- **DevOps teams**: The golden rules in DevOps are you build it, you run it and you break it, you fix it. This doesn't mean that there are no other rules – or better, guidelines – in DevOps. An enterprise seeks consistency in DevOps, across all the different teams. These teams will function autonomously, but they still need to adhere to centralized guidelines that are set at the enterprise level and implemented and managed at the value stream level. Why is this important? An enterprise service or product may very well be built by building blocks that are the responsibility of different DevOps teams. If the way of working, guidelines, and security guardrails are not aligned, the end product or service will most likely not meet the quality standards of the enterprise.

- **Enabling teams**: These are the teams that take care of the foundation; for example, the hosting platforms and the repositories that are used to store artifacts. DevOps engineers and developers will not need to worry about the network settings on the platform – these are arranged by the enabling team.

> **Note**
>
> This is an introduction to the full model. Please refer to the work of enterprise architect Rob Akershoek (@robakershoek), who made extensive contributions to the IT4IT framework. One of his books is listed in the *Further reading* section.

This section was about governance and controlling the processes in DevSecOps. In the next section, we will discuss how security teams and specialists can identify events that might impact how code is developed and deployed. A good understanding of and knowing how to work with threat modeling is a requirement.

Understanding and working with threat modeling

In the previous section, we discussed the governance of security in the enterprise and how it's integrated as DevSecOps. In this section, we will learn how security issues can impact the SDLC. When it comes to integrating security in DevOps, you need to have a good understanding of threat modeling, which provides us with information on how security threats may affect how software code is developed and deployed. We'll start by explaining what threat modeling is by looking at the definition of **The Open Web Application Security Project (OWASP)**. OWASP is an online community that provides insights into security threats, tools, and technology.

In essence, a threat model shows how security threats could impact the integrity of an application. The model assembles and analyzes security data and helps in making decisions on how to protect the application, thus improving the security of code and the hosting environment, by assessing the requirements, revisiting the design, and implementing improved security policies.

What is a threat? This is anything that impacts the application negatively and causes a failure or unwanted events, such as data leaks. Threat modeling identifies the possible vulnerabilities and then defines the mitigating actions. For identification, it can use the MITRE ATT&CK framework, which we discussed in the previous chapter. Threat modeling is more than just detecting security issues, or even preventing or fixing them.

Modeling is a structured, planned, and repeated activity that's used to continuously assess the environments and possible vulnerabilities, and then helps implement a structured approach to mitigate these vulnerabilities. Hence, threat modeling is something that needs to be conducted throughout the entire SDL, in which the model and the subsequent actions that are triggered by the model are continuously reviewed and refined. The reason for this is that during development, new features are added and maybe even new technologies. With that, there's a chance that new vulnerabilities – threats – will be introduced, so the model needs to be constantly revised.

Threat modeling typically involves the following steps:

1. **Scope analysis**: What is the scope of our threat model and analysis? Think of application code, programming language (for example C#, Python, or Java), APIs, virtual machines, platforms (think of cloud platforms such as AWS or Azure), database engines (for example SQL, Postgres, or Cassandra), and databases.

2. **Identify threat agents**: In MITRE ATT&CK, we focus on the vulnerabilities and techniques in which these have been exploited. In threat modeling, we also identify who might be interested in an attack. OWASP calls these threat agents: these can be both internal and external. The reason for making this clear distinction is to identify if environments are *fault-tolerant* for insiders or easy to breach from the outside.

3. **Assess mitigations**: OWASP refers to this as countermeasures. Known mitigations – for instance, the possibility to enroll patches – must be included in the model.

4. **Assess vulnerabilities**: If you know the scope, the possible attackers, and have identified possible countermeasures, then we can start analyzing vulnerabilities. Are these in the scope, who can they be exploited by, and what specific mitigating actions can be taken to prevent or minimize the impact?

5. **Prioritize risks**: Analyze the likelihood that vulnerabilities are being exploited and what the actual damage would be. This defines the priority of the threat.

6. **Execute mitigation**: Based on the priority, mitigating actions must be set in action to reduce the risks that have been identified.

> Tip
> The OWASP community manages its own GitHub pages for threat modeling. These can be found at `https://github.com/OWASP/www-community/blob/master/pages/Threat_Modeling.md`.

There are more models that define threat modeling, but in essence, they all come down to the same principles: assess, identify the threats, rate the threats, and identify risk-reducing actions.

We have security under governance, and we know how to assess our environments in terms of possible threats. The next step is to have this automated and integrated into the DevOps tooling. If we have done that, then we can say that we've implemented DevSecOps.

Integrating tools and automation

Throughout this book, we've discussed the importance of testing a couple of times. DevOps advocates testing at every single stage in the life cycle, from development to deployment. This includes security testing. But how can we achieve this continuous integration? The goal is to have tests running at developer check-in, while they pull code from repositories, during the builds, and during the actual deployments, including staging.

Let's look at **continuous integration** (**CI**) first. Developers will frequently do check-ins on code; in some cases, this can be up to several builds per day. That's the aim of CI and the agile way of working in DevOps: developers don't work on huge programs anymore; instead, they apply small iterations of code builds, adding one feature at a time. This way, it's easier to track changes in the code and, importantly, roll back if the addition is causing failures.

CI is about integrating these changes and additions into the code. The developer checks in, modifies the code, and releases it to the feature branch. From there, it's pushed to production and back into the repository, where the source code gets updated with that small new chunk of code, all in a very short timeframe. Since these are small iterations of the code, integrating that code and implementing it in production is easier than doing so for big releases. However, this only works well if the code is tested along the way. All tests must be applied to every new build and to every modification of the build.

In DevOps pipelines, we run these tests in an automated way. This requires integrated tooling to package the code, running the scripts to launch the appropriate infrastructure, applying the security policies, launching monitoring, and executing the tests.

In conclusion, testing is done throughout the entire development and deployment cycle. The following diagram shows what we mean by this, while also applying the principle of shift left:

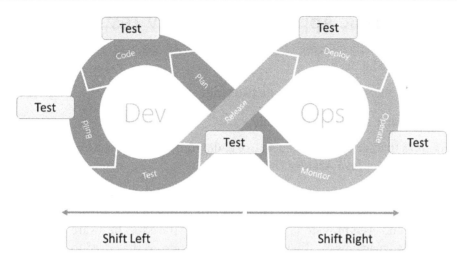

Figure 14.4 – Applying shift left in security testing

This includes security testing. After all, the goal is to integrate this with the DevOps tools and automate the CI/CD pipelines. We run security tests at the following instances:

- **Code Repository**: Checking in and pulling source code

- **Build**: Writing, modifying, and compiling the code

- **Pre-release or staging**: Production-like testing before the actual deployment to production and the master branch

In *Chapter 12*, *Architecting for DevSecOps*, we discussed the various types of scanning that we can apply to pipelines, such as **Static Application Security Testing** (**SAST**) and **Dynamic Application Security Testing** (**DAST**).

All of these tests can and should be automated. For a lot of enterprises, this is a true paradigm shift. Many security engineers and managers will persist in manual testing, rather than automating security testing. The problem is that security testing will become a bottleneck in DevOps: deployment is halted or, at best, slowed down by security testing. That's the reason why we want to integrate and automate security testing into our pipelines and at every stage of the SDLC.

Next, with automated and fully integrated security testing, we ensure that developers receive immediate feedback on vulnerabilities in the code. This will definitely improve the code.

Integrating Static Application Security Testing

SAST is crucial. We briefly discussed this in *Chapter 12, Architecting for DevSecOps*, so now, we will dive in a bit deeper.

First, we need to understand that there are two types of SAST:

- SAST tools that scan raw source code.
- SAST tools that scan decompiled source code from libraries, such as **Dynamic-Link Library (DLL)** or **Java Archive (JAR)**.

SAST tools scan the code line by line. They report where a potential vulnerability has been found in the code so that the developer knows exactly where to look. Most tools also rate the vulnerability and even provide suggestions for fixing the issue. Be aware that SAST tools need to be language-aware. They scan the source code, so they need to understand the language that the code is written in. Multiple SAST tools might be required if multiple languages are being used.

SAST tools are integrated into the CI/CD pipeline to enable scanning throughout the development process. Most tools identify security issues that are commonly reported and listed, such as those by OWASP. Every year, OWASP releases the top 10 web application security risks. Currently, the top 10 lists, among others, code injection, sensitive data exposure, and insufficient monitoring.

> **Tip**
> The OWASP top 10 can be found at `https://owasp.org/www-project-top-ten/`.

We will go back to monitoring in the last section of this chapter.

Integrating Dynamic Application Security Testing

DAST doesn't scan code. Simply put, DAST tooling simulates attacks. In *Chapter 13, Working with DevSecOps Using Industry Security Frameworks*, we discussed the MITRE ATT&CK framework, which lists techniques that help exploit vulnerabilities. DAST tools run these techniques by injecting malicious code strings or by brute force. By doing this, DAST tries to identify vulnerabilities in the functionality of the application, rather than in the source code.

DAST is sophisticated, which means it's costly. It runs transactions through the application and with that, it's also depending on components that interact with the application. For example, you could think of the frontend application and the database in the backend. DAST tooling needs to be configured well to be effective.

On that last note, DAST comes close to penetration testing. Most security officers and engineers will rely on manual penetration testing to detect vulnerabilities in integrations between different application stacks and services, especially when these stacks and services are used by different platforms and providers. In modern IT, where systems consist of **Infrastructure-as-a-Service (IaaS)**, **Platform-as-a-Service (PaaS)**, and **Software-as-a-Service (SaaS)** hosted on various platforms such as AWS, Azure, or in private data centers, integrating and testing integration points becomes more and more relevant.

Please be aware that public cloud platforms have strict policies for penetration testing. As an example, AWS only supports this for a very limited number of services, such as their compute platform EC2. Violating these policies might inflict penalties or, in the worst case, a ban on the usage of services.

Integrating using CircleCI orbs

A rather new phenomenon in the world of DevSecOps integration, but worth mentioning here, is orbs – reusable snippets of YAML code that take care of repeatable actions, such as security scanning. Orbs allow for direct integration with popular security scanning tools. It's a concept of CircleCI, a San Francisco-based company that delivers tooling to automate CI.

It claims out-of-the-box integrations for these tools into CI/CD pipelines. As an example, orbs for Probely (a web vulnerability scanner) and SonarCloud (code analysis) are available, but developers can also create their own orbs and push these to the open source Orb Registry.

Now that we've covered the integrated security frameworks and tooling, we have to make sure that we keep track of the outcomes. In the last section, we will discuss monitoring in more detail.

Implementing monitoring

A crucial element of security is ensuring that the necessary security policies are in place and that the environments are indeed protected. This may sound simple, but it requires proper configuration of monitoring. Developers need information to help them fix bugs in the first place but also to improve the code and with that, the application. This applies to customer experience and performance, but also to ensuring that the application remains protected. Hackers don't sit on their hands: they constantly find out new ways of attacking systems. Hence, we need to constantly monitor what happens to and inside an application.

Security monitoring is not only about detecting unexpected behavior. It's about analyzing all behavior. This provides insights to developers to help them improve their code. For that, monitoring needs to facilitate three main services:

- Collect

- Analyze

- Alert

Sometimes, storage and visualization are added to these services. However, storing monitoring data is more about logging, while visualization is about comprehensively presenting monitoring data. These are important services, but they are not core to monitoring itself. Of course, you will need methodologies to receive and view alerts. As an example, Grafana is a popular tool that provides cross-platform dashboards that allow us to query and visualize stored monitoring data.

Before we get to implementing monitoring, the architect needs to assess the following four W questions:

- *What* are we monitoring?

- *Why* are we monitoring *that* (and *why* not anything else)?

- *When* are we monitoring *that*?

- *Who*'s paying attention (who needs to be informed)?

The *what* is about collecting data, the *why* is triggering analyses, and the *who* is about sending alerts to the right people or entities within the enterprise. As we mentioned previously, monitoring properly is crucial to the feedback loop that we have in DevOps, including alerts regarding security events that may impact the systems. It's a big misunderstanding that systems that are code only and hosted on cloud platforms are monitored and secured by default. They are not. Platforms such as AWS, Azure, and Google Cloud merely provided huge toolboxes that you can use to secure code that you run on the platforms. Engineers will need to configure the monitoring themselves. This includes monitoring the infrastructure, such as the health statuses of the **virtual machines** (**VMs**) that are used in the cloud. A certain increase in the usage of the VM – an unexplainable peak in the CPU or memory of the VM – might indicate that something wrong.

Next, to be useful for DevOps, monitoring must collect metrics from the application itself. This way, monitoring provides information on the running state of the application. Applications must be enabled to provide this information to monitoring systems. This information is typically labeled in the code of the application. Examples include labels such as **debug** and **warning**. Monitoring will pick up this information and make sure that it gets to the engineers in a comprehensible format.

How do monitoring tools do that? There are a couple of ways it does this, but in most cases, tools use either agents that collect data on systems or they simulate – agentless – transactions. With transactions, tools will send a transaction to the application and analyze the response if and when the transaction is returned.

> **Note**
> Transaction-based monitoring doesn't mean that these tools can't or won't use agents. There are agent and agentless systems. Some enterprises have policies that state that agents on their systems are prohibited because of overhead on systems, or because enterprises fear the intrusiveness of agents.

In the case of monitoring agents, these can and should be part of the desired state configuration of the application when it's deployed to the platform. Remember shift-left: this should be integrated into the entire DevOps chain so that monitoring starts from the moment code is pushed to the development, test, and staging systems. In other words, monitoring is not just for production.

Because of the capabilities that monitoring tools must have, these tools can become very vast and complicated. Agents might lead to some so-called overhead on systems, meaning that the agents will increase their use of resources in systems or trigger more network traffic. These aspects should be taken into architectural consideration. Again, start with the four W questions.

Lastly, enterprises might not end up with just one monitoring tool, but with a chain of tools. They might use AWS CloudWatch or Azure Monitor to monitor the cloud resources, Prometheus to collect real-time data in highly dynamic systems such as container platforms, and Sysdig for cloud-native applications. Specifically for security monitoring, enterprises can use Splunk Security Essentials or the cloud-native Azure Sentinel. Be aware that a lot of security tooling focuses on network and identity and access to systems, and not so much on the applications or application code. That's why enterprises end up with monitoring toolchains: it's for a good reason. Architects have a big say in what tools fulfill the needs of the enterprise.

> **Note**
>
> We will only mention a couple of tools here. This is by no means meant to be exhaustive or to promote specific tools over other tools. The tools that are mentioned here are widely used in enterprises, but there are many other, great tools available.

Engineers probably don't want to use several consoles to watch the output of these different tools. Overarching enterprise suites such as ServiceNow and BMC Helix provide broker platforms that enable the single pane of glass view: various monitoring tools and data collection processes can be aggregated in this one suite. These are complex systems that require highly skilled professionals to implement and configure, but in a world where IT is becoming more complicated by the day, the investment is worthwhile. Regarding DevOps teams, remember that they are fully responsible for developing, deploying, and operating their code. You might not want to rely on these "monolith" types of overall management systems, but in enterprises with a wide variety of products and services, it's essential to have a complete overview of all the assets, developments, and the state of these deployments.

So, we have adhered to security frameworks, defined the policies, and integrated the tools in our DevOps and CI/CD cycle, but are we really good to go? IT is changing rapidly and becoming more challenging every day. Attacks and breaches are in the news every day. This has a big impact on how we secure our systems. We can't trust anyone in our enterprise networks anymore. On the other hand, we want to give developers as much freedom as possible so that we get the best results in terms of coding. After all, the first statement in DevOps is about trust: you build, you run it – because you can. But how do we deal with trust? That's the final topic of this book: zero trust architectures and how they impact DevOps.

Summary

DevOps and DevSecOps are not separate things: security must be fully integrated with DevOps. In this chapter, we discussed how we integrate security in DevOps, not only focusing on scanning tools but mainly on governance, applying threat modeling, and monitoring our DevOps environments. For governance, we looked at the principles of GRC that allow enterprises to manage uncertainties – such as security risks – while defining strategies to achieve their business goals. This is the foundational step to integrating security into all the layers of the enterprise and with that, the development of products and services.

To detect, recognize, and counterfeit attacks, we need to work with threat modeling. In this chapter, we discussed OWASP, which provides insights into how security events can impact businesses. Next, we look at security scanning in a more detailed way. SAST and DAST are necessities in DevSecOps.

In the last section, we learned about the various steps that an architect needs to take to implement monitoring. They need to ask themselves four basic questions: what are we monitoring?, why are we monitoring that?, when do we monitor that?, and who needs to be informed? We also looked at the characteristics of monitoring tools.

Security is about trust and in modern IT, with an increasing number and variety of attacks, the basic rule is that enterprises can't trust anyone or anything anymore. This leads to a specific area in architectures: zero trust. This is the topic of the final chapter of this book.

Questions

1. True or False: In OWASP, threat agents can be both internal and external.

2. Name the two types of SAST tooling.

3. What are the three main functions of monitoring?

Further reading

- *IT4IT for managing the business of IT*, Rob Akershoek et al., The Open Group.

- Blogpost on DevOps.com by Mike Vizard on the introduction of private orbs, 2021: `https://devops.com/circleci-adds-private-orbs-to-devops-toolchain/#:~:text=Constructing%20an%20orb%20gives%20DevOps%20teams%20a%20relatively,of%20the%20DevOps%20tea-m%20can%20more%20easily%20consume.`

- Article on ITSecuroty.org with in-depth insights into security testing and integration, by Adrian Lane, 2019: `https://itsecurity.org/enterprise-devsecops-security-test-integration-and-tooling/`.

- An overview of popular security tools in DevOps: `https://dzone.com/articles/an-overview-of-security-testing-tools-in-devops`.

- *Hands-On Security in DevOps*, by Tony Hsiang-Chih Hsu, Packt Publishing, 2018.

15

Implementing Zero Trust Architecture

Digital transformation is the new paradigm in enterprises. Enterprises are adopting data-driven architectures and using more and more native services in the cloud and, through this, accelerating the development of their products and services. Under this pressure, security has to keep up and be sure that environments, in a lot of cases even mission-critical environments, remain resilient. This is the domain of zero trust.

This chapter explains what zero trust is and why it is important to DevOps. Zero trust assumes that everything is secured inside a corporate network and that includes the DevOps pipelines. Some of the technologies used in zero trust environments are service meshes and microservices, a topic that we will discuss in the final section of this chapter.

After completing this chapter, you will have learned what zero trust means and the impact it has on DevOps. You will have learned how microservices and secure service mesh drive secure digital transformation. In the final section, we will briefly discuss some solutions that are available from cloud platforms.

In this chapter, we're going to cover the following main topics:

- Understanding zero trust principles
- Architecting for zero trust security
- Including microservices in architecture
- Integrating zero trust in pipelines

Understanding zero trust principles

Zero trust really means zero trust, for starters. The principles of zero trust have gained a lot of traction in IT security over the past few years, and for a good reason. Attacks don't just come from the outside, but also from the internal networks in enterprises. Zero trust advocates that any user, or maybe every identity, is authenticated, regardless of whether the user is inside or outside the enterprise's network. When authenticated, the user must be validated against security policies and authorized before access to applications is granted. Data access should only be granted through verified applications to which users are authenticated and authorized.

Before we learn how this would work in DevSecOps, and particularly in **Continuous Integration/Continuous Deployment (CI/CD)** pipelines, we need to have a deeper look at the principles of zero trust.

Zero trust starts with knowing who's in the enterprise's network. There's one important thing to note at this point: in the cloud, everything is an identity. It can be a real user, a person, but also a service that is triggered to execute a specific action. Also, services have certain rights: they are allowed to perform a specific action or fetch a defined dataset and are prohibited from taking other actions. Therefore, all identities, or more accurately accounts, must be known and, on top of that, it must be clear what rights they have. It means that an enterprise has to constantly monitor and validate all its accounts, along with their credentials and their rights. This must be done in real time.

Now, you might think that zero trust is mainly about monitoring accounts. But there's more. Zero trust also implies that an enterprise has put measures in place to prevent authenticated users from doing more than they are authorized to. You may be thinking of setting least privileges to accounts, but you also need to consider network segmentation and restricting specific protocols on networks. Basically, you need to consider everything that contains an account so it can only perform the tasks it's authorized to do in the place where the account is authorized. This must be enforced by strong **Identity and Access Management (IAM)** policies, network segmentation, external and internal firewalls, gateways, and strict routing policies such as *deny all* and *allow only* whitelisted addresses.

Principles that must be included in zero trust are as follows:

- Account types and credentials are always based on least privilege .

- Specified privileged rights and rules of the application of these rights.

- Defined endpoints for services and applications.

- Authentication protocols.

- Security monitoring includes intrusion detection, intrusion prevention, and anomaly detection.

- Operating system hardening with the latest versions and the most recent patches.

- Software life cycle with the most recent versions and the most recent patches.

How does this affect DevOps? The answer to that question is: zero trust has a huge impact on DevOps and the agile way of working. DevOps is all about gaining speed in the development and deployment of application code. This requires flexibility and a great deal of responsibility for the DevOps teams. It's true that very strict security rules can hinder the speedy process of development and deployment. Yet there's no other way to protect the assets of the enterprise. DevOps teams also have a responsibility in protecting these assets.

The consequence is that DevOps teams must adhere to zero trust too. Teams can only use accounts that are allowed to enter the code repositories, work with builds that are contained in a specific segment of the enterprise network in the cloud, use only approved operating systems, software, and tools, and apply security policies that are enforced by routing and firewall rules.

Zero trust doesn't mean that the DevOps process is slowed down by default, though. That only happens if responsibility for applying zero trust is placed outside the teams. For example: the team has code ready for deployment, but now has to wait for a specific firewall port to be opened. That can be done quickly if the port is already whitelisted, and automated security scans have verified that the code is compliant with the firewall rules. If the approval has to go through a security department that needs to validate everything manually, then it will slow down the process heavily.

Hence, we need to include zero trust in DevOps. We will discuss this in the upcoming sections.

Architecting for zero trust security

With a good understanding of the concept of zero trust, we can define architectures that apply the principles of zero trust. The following guidelines will help define the architecture. Some of these principles might be obvious, and others may lead to constraints in the way developers develop and deploy applications. But, at the end of the day, we need to be sure that the enterprise assets are secured:

- Assess and analyze all access controls. Strict policies on IAM must be in place. Least privilege must be part of those policies. This is the backbone of zero trust according to the **National Institute of Standards and Technology** (**NIST**). They defined a set of principles for zero trust architectures, all involving the way enterprises handle IAM. The key principle is to have a single source of identities. In most cases, enterprises will use **Active Directory** (**AD**) for this. In short, any user or identity must be *known* by the AD.

- Next, there must strong authentication. Is the identity really who it's claiming to be? **Multi-factor authentication** (**MFA**) is strongly recommended. NIST also stresses the need to verify and validate the context in which users are authorized and authenticated. For example: from which machine is a repository accessed, and is the device compliant with the enterprise's standards? A lot of developers have their own machines with their own preferred tools. This must be assessed to clarify whether this is compliant with the security policies.

- Specific access policies to applications must be defined and controlled. A developer working on a marketing website likely won't need access to the application that controls the supply chain of an enterprise. In that case, access to that application should be restricted. Zero trust therefore means that every application has its own set of policies: who is eligible to access it, to what level, and what are the rights in that application?

- Data classification and data security are the next building blocks for zero trust architecture. Data must be protected. The challenge in modern, cloud-based IT is that data can be anywhere, and it's shared across platforms, applications, and services. Enterprises need to know exactly where their data is, what type of data it is, and who or what is allowed to access it under strict conditions. Data must be identified and classified: is it, for instance, confidential, or may it be publicly accessed? Strict privacy regulations, such as the **General Data Protection Regulation (GDPR)** in the European Union, are guidelines for classifying data; applying these guidelines is the responsibility of the enterprise.

- NIST and the **National Cybersecurity Center of Excellence (NCCoE)** also define the *trusted cloud* as a building block. That's because of the dynamic character that clouds have by default. Now we are really at the heart of DevOps, where we go by the rule of autonomous working teams that can spin up environments the instant they need them, modify them, and even delete them. These environments will use data, but some of these environments may only be short-lived, while others will eventually be pushed to production. Cloud technology, where everything is code, facilitates this. This is a huge challenge for security, especially in keeping environments consistent with security policies. Hence, security must be embedded in DevOps. Monitoring should be in real time and enable controls to identify any breach of the security policies, even if it does mean that development is halted by that.

In summary, we could say that zero trust is mainly about separating network segments, applications, data, and services as much as possible and only allowing access to these different components to authenticated and authorized users with least privileges. Microservices will help the architects achieve this. However, microservices do come with challenges. These challenges can be overcome with service meshes.

Including microservices in architecture

DevOps is about gaining higher productivity with faster releases of code. DevOps teams can focus on specific tasks and code that is designed to only perform that task. They develop the code independently from other services to increase focus, the speed of delivery, and customer experience. Security principles are applied to these services and continuously validated by the means of automated scanning. DevOps is by default distributed architecture, in contrast with monolithic architectures where systems are designed and built as a whole. In DevOps, the architecture will be driven by microservices: an application is defined as a collection of independent services that will communicate with each other over specified protocols. The following figure shows the principle of microservices:

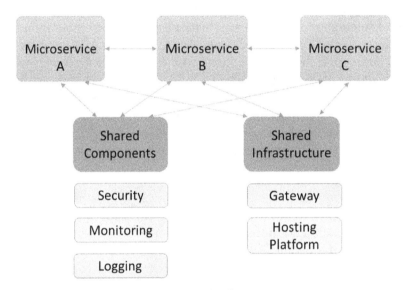

Figure 15.1 – Principle of microservices

In terms of security, we can assume that microservice architectures are more secure than monolith systems. If one of the services is breached, it doesn't automatically mean that the whole application stack is breached, as long as the affected service is contained well enough. Unfortunately, it's not as simple as that. The reason for that is that microservices do need to be able to communicate with each other. The next question is: how can we enable that in a secure way? The answer is a service mesh.

First, let's look at the best practices for microservices architecture:

- **Defense strategy**: Microservices allow various defense layers or security tiers. As an example, a web portal needs to be publicly accessible, but the application and the data should be protected. A good example is mobile banking apps. The app is accessible on any smartphone: the user can download it from an app store and install it on the phone. To access the application that retrieves and presents account information, the user will need to have several things: an account with that specific bank and an account that allows them to use the mobile app. These are two separate things. Obviously, the account data is protected too, for instance, by encryption.

- **DevSecOps**: As we have seen in the previous chapters, this is all about embedding security practices into DevOps. Code is scanned automatically during the entire build against policies and industry security and compliance frameworks. But it's not only during the build; post-deployment, applications and code should continuously be monitored for vulnerabilities.

- **MFA**: Every application should be accessed only with MFA. A username and a password are simply not sufficient; authentication should be done using a second factor, for instance, using an authentication app on a different device than the device someone uses to log in. Even when MFA is already used to access an application, re-authentication might be desirable when specific, highly confidential data is accessed from that application. Having access to an application doesn't mean by default that a user should have access to all the data that can be retrieved by that application. Applications and data are separate tiers or layers.

- **Dependencies**: In cloud environments, we will likely use cloud services such as **Platform as a Service** (**PaaS**) and **Software as a Service** (**SaaS**). We will need **application programming interfaces** (**APIs**) to enable interaction between these services. These are dependencies and they might lead to vulnerabilities and security threats if not verified and configured well. Source code must be scanned for vulnerable dependencies.

Dependencies are probably the biggest challenge in terms of security. How do we deal with that in modern architecture, using microservices?

Understanding and applying a service mesh

DevOps is served well by microservices. It's the perfect way to develop and deploy new features into code without affecting other, running services. Because of the granularity of microservices, development and deployment can also be secured at a low level, resulting in a low risk of services being disrupted for the users. Using microservices means that misconfigurations or badly programmed implementations are minimized to only specific services that are being developed and deployed, also minimizing the attack surface of the entire application stack. To enable this way of working, containers play a major role. Services and features are wrapped and deployed as containers.

The next challenge is to have these containerized services and features interact with each other securely. That's what a service mesh is about. To establish the interaction, developers need to configure these within the application code. They will integrate libraries that can communicate with services outside the application, such as service discovery, load balancing, and setting up internal **Transport Layer Security** (**TLS**) traffic to other services. First of all, the configuration strings and the services they call from the application code need to share a common language. But more importantly, when a service changes, it needs to be adjusted in the application code as well. This makes the application code complex.

A service mesh tackles this problem by removing the complexity from the application and moving it to a service proxy. This proxy now takes care of a lot of *third-party services* that applications use to interact with other functional components. Think of traffic management including load balancing, authentication, and of course, security and monitoring. The services are now abstracted from the application code as a separate component.

Developers will only have to worry about the application code since all other services are taken care of by the service proxy. With this, we have strict segregation of responsibilities.

That sounds like a good solution, but how does it work in practice? We will learn that in the last section of this chapter.

Integrating zero trust in pipelines

In the previous sections, we discussed the principles of zero trust architectures and how microservices can help us with zero trust. Next, we learned how we can have microservices interact by means of a secure service mesh. In this section, we will learn how we can achieve this with containerized applications and using cloud services that we target from CI/CD pipelines. Platforms such as AWS and Azure offer solutions for this, and we will discuss these solutions.

First, we need to understand how we add security to a service mesh. One way to do this is with sidecars. Explained in a very simple way, a sidecar is a point in a container cluster where security postures are inserted. You could envisage it as a main road where cars are driving. A car carrying specific security policies comes from a side road and inserts itself in the line of cars on the main road. However, the point where this happens is fixed.

There are various tools that offer a sidecar service mesh. Popular ones are Istio, Linkerd, and Envoy. What these tools have in common is that they put the desired functionality in a separate container that is inserted close to the application containers, just like we described with inserting cars. Since most developers that work with containers work with Kubernetes, it's important to know that the sidecar containers have to be placed in the same Kubernetes pod as the application containers. This is because the namespace of the pods needs to be the same. The application containers and the sidecars can be integrated from the CI/CD pipeline.

The whole principle of a service mesh and sidecar proxies is shown in the following figure:

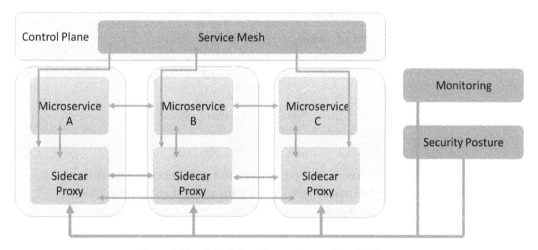

Figure 15.2 – Principles of a service mesh and sidecars

As noted, cloud platforms offer service meshes as well. AWS has AWS App Mesh, which allows services to interact with each other regardless of the underlying infrastructure, when it uses the Envoy sidecar proxy. Native App Mesh works with the serverless infrastructure services of AWS Fargate, the compute engine EC2, and the container orchestration services of **Elastic Container Services (ECS)** and **Elastic Kubernetes Services (EKS)**. The high-level architecture of AWS App Mesh is shown in the following figure:

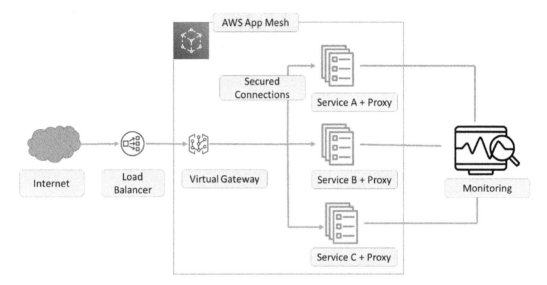

Figure 15.3 – Architecture of AWS App Mesh

In Azure, we work with Azure Service Fabric, Microsoft's container orchestrator for deploying and managing microservices. The fully managed mesh service called Azure Service Fabric Mesh that was launched in 2018 has been retired by Microsoft from April 2021. Companies that use Azure can use Azure Container Services, **Azure Kubernetes Services (AKS)**, or Azure Service Fabric managed clusters to create the mesh functionality. The principles of Azure Service Fabric are shown in the following figure:

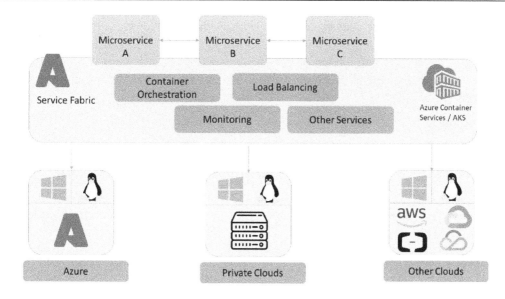

Figure 15.4 – High-level architecture of Azure Service Fabric

This concludes our journey through enterprise DevOps, AIOps, and DevSecOps. In this age of digital transformation, architects have a big task ahead of understanding how these methodologies help enterprises in modernizing their IT environments, becoming more agile in software development, while ensuring maximum security during development and deployment. This book is just a starting point. The proof of the pudding is in the eating, so go out and try to make it work.

Summary

In this chapter, we first studied the principles of zero trust architecture, and we learned that DevOps teams need to adhere to these principles too. Zero trust starts by knowing exactly who may access code repositories, and knowing that builds can only be deployed to strictly contained network segments so that other services are not impacted. Next, we learned that microservices architecture can serve DevOps really well. They allow independent development and deployment of features in code without affecting other services.

We learned that microservices are a secure type of architecture. The challenge, however, is to establish interaction between these microservices. We studied service mesh as a solution for that and learned how to integrate security postures as a containerized microservice, using the technology of sidecar proxies. We learned that sidecars can be used to insert secure services and monitoring next to our microservices.

In the final section, we introduced some mesh services that are offered by the cloud providers Azure and AWS. This concluded the journey through enterprise architecture for DevOps, DevSecOps, and AIOps, all of which are becoming increasingly important to understand and to implement, eventually to successfully drive digital transformation in enterprises.

Questions

1. What basic rule do we apply with respect to the privileges of accounts in a zero trust environment?

2. What type of service do we use to insert separate containers with security postures next to application containers?

3. What does AWS offer to enable service mesh?

Further reading

- Website of the **National Cybersecurity Center of Excellence (NCCoE)** on zero trust architecture: `https://www.nccoe.nist.gov/projects/building-blocks/zero-trust-architecture`

- *Hands-On Microservices with Kubernetes*, by Gigi Sayfan, Packt Publishing, 2019

- Documentation on Microsoft Azure Service Fabric: `https://docs.microsoft.com/en-us/azure/service-fabric/service-fabric-overview#:~:text=%20Overview%20of%20Azure%20Service%20Fabric%20%201,application%20lifecycle...%204%20Next%20steps.%20%20More%20`

- Blog post on AWS App Mesh: `https://aws.amazon.com/app-mesh/?aws-app-mesh-blogs.sort-by=item.additionalFields.createdDate&aws-app-mesh-blogs.sort-order=desc&whats-new-cards.sort-by=item.additionalFields.postDateTime&whats-new-cards.sort-order=desc`

Assessments

Chapter 1

1. True
2. Integration, collaboration, configuration management
3. Continuous Integration, Continuous Delivery (sometimes Continuous Deployment)
4. Deployment failure detection time

Chapter 2

1. False. Business benefits – and with that the business case – are an important asset of demand management.
2. Static analysis
3. Development – Test – Acceptance – Production

Chapter 3

1. Test of separate components
2. Boundary value analysis
3. Definition of Done
4. True

Chapter 4

1. Refactor
2. AKS and EKS
3. BIA

Chapter 5

1. Toil

2. Time to Detect and Time to Repair

3. The consequences of the risk are transferred, for instance to an insurance company.

Chapter 6

1. These components must be part of the technology architecture:

   - Production scheduling/monitoring

   - System monitoring

   - Performance monitoring

   - Network monitoring

   - Event management (incidents, problems, changes)

2. The four value streams that IT4IT defines for IT delivery are as follows:

   - Plan: strategy to portfolio

   - Build: requirement to deploy

   - Deliver: requirement to fulfill

   - Run: detect to correct

3. Microservices

4. Level 3: proactive

Chapter 7

1. Empathy

2. CodeBuild, CodePipeline, and CodeDeploy

3. Mean time to acknowledge

Chapter 8

1. The presentation tier has two main functions. First, it helps the user to put in the request in a comprehensible way. Once the request has been processed, the response is presented in this tier.

2. Anomaly detection

3. Engagement data

4. True

Chapter 9

1. Reducing TCO/cost reduction by, for example, moving systems and data to other platforms

2. Kubernetes

3. The possible outcomes/results of AI-enabled DevOps, specifically for the improvement of code:

   - Identifying missing code

   - Detecting badly written code

   - Detecting unnecessary code

   - Detecting expected and/or required missing dependencies

Chapter 10

1. Auto-healing

2. Deduction

3. True

Chapter 11

1. The four principles are as follows:

   - Prevention

   - Detection

   - Correction

   - Direction

2. Docker Notary

3. False

Chapter 12

1. SCA will detect dependencies in code.

2. Linting.

3. CloudFormation.

4. **Elastic Kubernetes Services (EKS)** in AWS, **Azure Kubernetes Services (AKS)** in Azure, and **Google Kubernetes Engine (GKE)** in GCP.

Chapter 13

1. ISO 27017.

2. Container administration command and deploy the container.

3. False – it's a generic control to set the latest version of Docker.

Chapter 14

1. True

2. SAST tools that scan raw source code and tools that scan decompiled source code from libraries

3. Collect, analyze, alert

Chapter 15

1. Least privilege

2. Sidecars

3. AWS App Mesh

`Packt.com`

Subscribe to our online digital library for full access to over 7,000 books and videos, as well as industry leading tools to help you plan your personal development and advance your career. For more information, please visit our website.

Why subscribe?

- Spend less time learning and more time coding with practical eBooks and Videos from over 4,000 industry professionals

- Improve your learning with Skill Plans built especially for you

- Get a free eBook or video every month

- Fully searchable for easy access to vital information

- Copy and paste, print, and bookmark content

Did you know that Packt offers eBook versions of every book published, with PDF and ePub files available? You can upgrade to the eBook version at `packt.com` and as a print book customer, you are entitled to a discount on the eBook copy. Get in touch with us at `customercare@packtpub.com` for more details.

At `www.packt.com`, you can also read a collection of free technical articles, sign up for a range of free newsletters, and receive exclusive discounts and offers on Packt books and eBooks.

Other Books You May Enjoy

If you enjoyed this book, you may be interested in these other books by Packt:

DevOps Adoption Strategies: Principles, Processes, Tools, and Trends

Martyn Coupland

ISBN: 9781801076326

- Understand the importance of culture in DevOps
- Build, foster, and develop a successful DevOps culture
- Discover how to implement a successful DevOps framework
- Measure and define the success of DevOps transformation
- Get to grips with techniques for continuous feedback and iterate process changes
- Discover the tooling used in different stages of the DevOps life cycle

Learning DevOps

Mikael Krief

ISBN: 9781838642730

- Become well versed with DevOps culture and its practices
- Use Terraform and Packer for cloud infrastructure provisioning
- Implement Ansible for infrastructure configuration
- Use basic Git commands and understand the Git flow process
- Build a DevOps pipeline with Jenkins, Azure Pipelines, and GitLab CI
- Containerize your applications with Docker and Kubernetes
- Check application quality with SonarQube and Postman
- Protect DevOps processes and applications using DevSecOps tools

Packt is searching for authors like you

If you're interested in becoming an author for Packt, please visit `authors.packtpub.com` and apply today. We have worked with thousands of developers and tech professionals, just like you, to help them share their insight with the global tech community. You can make a general application, apply for a specific hot topic that we are recruiting an author for, or submit your own idea.

Share Your Thoughts

Now you've finished *Enterprise DevOps for Architects*, we'd love to hear your thoughts! Scan the QR code below to go straight to the Amazon review page for this book and share your feedback or leave a review on the site that you purchased it from.

https://packt.link/r/1801812152

Your review is important to us and the tech community and will help us make sure we're delivering excellent quality content.

Index

H

I

V

Value, Objectives, Indicators, Confidence,
    and Experience (VOICE) model
  working with 22
value streams 225
version control system (VCS) 34
vertical or scale-up scaling 73
virtual machines (VMs) 64, 233
Virtual Private Cloud (VPC) 198

W

warning 233
worker 136

Z

zero trust
  architecting 240, 241
  integrating, in pipelines 244-246
  principles 238, 239

www.ingramcontent.com/pod-product-compliance
Lightning Source LLC
Chambersburg PA
CBHW060522060326
40690CB00017B/3352